# Electromagnetic Field Radiation in Matter

*Edited by Walter Gustavo Fano, Adrian Razzitte and Patricia Larocca*

Published in London, United Kingdom

IntechOpen

*Supporting open minds since 2005*

Electromagnetic Field Radiation in Matter
http://dx.doi.org/10.5772/intechopen.83257
Edited by Walter Gustavo Fano, Adrian Razzitte and Patricia Larocca

Contributors
Wiqar Hussain Shah, Waqas Khan, Walter Gustavo Fano, Leonid Chervinsky, Bratko Filipič, Lidija Gradisnik, Ferenc Somogyvari, Shigeru Tamaki, Shigeki Matsunaga, Masanobu Kusakabe, Emeka Ikpeazu, Kristine Kovacs, Hrvoje Mazija, Toth Sandor



First published in London, United Kingdom, 2020 by IntechOpen
IntechOpen is the global imprint of INTECHOPEN LIMITED, registered in England and Wales, registration number: 11086078, 7th floor, 10 Lower Thames Street, London, EC3R 6AF, United Kingdom
Printed in Croatia

British Library Cataloguing-in-Publication Data
A catalogue record for this book is available from the British Library

Additional hard and PDF copies can be obtained from orders@intechopen.com

Electromagnetic Field Radiation in Matter
Edited by Walter Gustavo Fano, Adrian Razzitte and Patricia Larocca
p. cm.
Print ISBN 978-1-78984-518-1
Online ISBN 978-1-78984-519-8
eBook (PDF) ISBN 978-1-83968-568-2

For EU product safety concerns:
IN TECH d.o.o., Prolaz Marije Krucifikse Kozulić 3, 51000 Rijeka, Croatia, info@intechopen.com or visit our website at intechopen.com.

# Meet the editors

Dr. Walter Gustavo Fano received his PhD in Engineering and Electronic Engineering from the University of Buenos Aires. He directs the FIUBA Electromagnetic Radiation Laboratory. He was a professor at the EST Army, ITBA, UNPSJB and currently at the University of Buenos Aires. He was co-author of 3 books, 1 book chapter and 5 book chapters in press. He has published many papers in magazines and conferences. He was president of the IEEE AP & EMC Soc. Chapter, president of the IEEE Gemccon 2016, president of the Congress of AP & EMC IEEE and UBA in 2013, and president of Advances in Antenna Test and Measurement IEEE 2011. He collaborated in the organization of numerous conferences and was a senior member of IEEE.

Adrián César Razzitte was born in Buenos Aires, Argentina. He obtained his BSc degree in Chemistry, his MSc degree in Physical Chemistry and his PhD in Physical Chemistry, all at the National University of La Plata. He is currently a Full Professor of Chemical Physics and Statistical Thermodynamics in the Chemistry Department of the Engineering Faculty of the Universidad de Buenos Aires. He was Director of that department from 2006 to 2017. He is Head of the research group of non-equilibrium thermodynamics. Professor Razzitte has had numerous research papers published in international journals and has participated in numerous international conferences on condensed matter, statistical thermodynamics and non-equilibrium thermodynamics. He has directed several PhD theses in engineering.

Patricia Larocca is a senior research scientist and the Director at the Institute of Applied Geodesy and Geophysics of the Faculty of Engineering of Buenos Aires University. She has a MSc degree in Physics received from the Faculty of Exact and Natural Sciences, and a PhD in Engineering received from Faculty of Engineering of Buenos Aires University. She has been an Associate Professor at the Facultad de Ingeniería for over ten years. In 2007, Dr. Larocca joined the Institute of Applied Geodesy and Geophysics and has since worked on multiple scientific topics related to space weather, from the effects of geomagnetic storms on power systems and pipelines to satellite environment and to space weather forecasts. She has been a leader and collaborator on multiple research projects dealing with impacts of space weather on pipelines, power grids, and also investigated the ionized radiation on satellite orbits. She is the author or co-author of more than 40 peer-reviewed scientific articles on geomagnetic and space weather effects topics, one chapter of a published book and one chapter of a book currently in press.

# Contents

# Preface

The purpose of this book is to study the interaction of electromagnetic waves, and the application of direct current signals in different media, with topics that have very important applications in science and technology today. The media where the interaction occurs are various, such as dispersive media, conductors, biological tissues of animals, and other media. The constitutive relationships that link the electric and magnetic fields with the densities of electric and magnetic flux are used, and the concepts of electrical conductivity and permittivity, electric field, magnetic field, voltage, power, energy, and heat are also covered.

It is recommended that the reader be a graduate of engineering, physics, or an equivalent subject, where they have dealt with the topics of mechanics, physics, heat, electromagnetism, and mathematical analysis, which make advanced study of the subjects essential. To understand this text, it is necessary to have knowledge of the laws of electromagnetism and the electromagnetism equations or Maxwell's equations.

The book consists of seven chapters that are interconnected by means of concepts and can be read independently.

In Chapter 1 we begin with the resolution of Maxwell's equations with adequate edge conditions to obtain the electric and magnetic fields, and the rest of the parameters of interest in an electromagnetic engineering problem. It is explained that the physical models of electrical permittivity and magnetic permeability must comply with the Kramers-Kronig equations, which obey the physical principle of causality.

Chapter 2 deals with electromagnetic propagation in various dispersive media that is of interest for its technological applications. The electromagnetic propagation in this case has different speeds for each wave excitation frequency, and there will also be attenuation to the amplitude of the electric and magnetic fields in the dispersive media, which will increase with frequency. The phenomena of plasmonic dispersion, dispersion in conductive media, modal dispersion, chromatic dispersion, and intramodal dispersion are explained.

Chapter 3 deals with the electrical properties of solids such as electrical permittivity and electrical conductivity, in the first part the fundamental concepts, the properties of transmission lines with losses, and their associated parameters such as characteristic impedance and propagation constant, are explained and speed of propagation, where the time domain and frequency methods are used, and finally experimental results are presented for the case of dry sand. It is observed that there are interesting applications to agriculture, geophysics, and engineering.

Thermoelectric properties of the Chalcogenide System are presented in Chapter 4.

The first part explains the Seebeck, Peltier thermoelectric phenomena, electrical conductivity and power factor. Electrical conductivity and Seebeck coefficient measurements are explained. Experimental results of the ternary and quaternary Tellurium Telluride chalcogenides, Tl10-x-yAxByTe6 nanoparticles, with different types of dopants (A = Sb, and B = Sn) and with different concentration of Sn are presented.

Chapter 5 discusses electrical conductivity of molten salts and ionic conduction in electrolyte solutions. A microscopic description for the partial DC conductivities in molten salts has been discussed, starting from a Langevin equation. The obtained results for concentration dependency of electrical conductivity are basically represented as a function of the square root of concentration. The electrophoretic effect and the relaxation effect are discussed from a microscopic view point.

In Chapter 6 we study the effects of ultraviolet radiation on the body of an animal.

In order to properly protect and control the effects of electromagnetic radiation on the human body, it is necessary to know and understand the process of absorption and conversion of electromagnetic radiation falling on the surface of the body. The material contains the original results of experimental studies electromagnetic radiation transmission through a sample of quasi-vital skin samples from pigs of different ages.

Chapter 7 covers the application of square-wave electric impulses of 10 ms and 100 V/cm of field force, produced by an impulse generator device, and the proliferation patterns of different animal cells. The discussion is about the influence of one or three square impulses with field force of 100 V/cm on different cells growing in a monolayer and the influence of one or three square impulses with a field force of 100 V/cm on the cells that grow in suspensions.

In this book, the authors aimed to provide material of important topics for the researcher, because novel experimental results are presented, or with a theoretical work, so that many people can apply these results. The applications that are found are diverse and we hope that they will be useful to the researchers in the field of engineering and sciences.

My sincere thanks to Dr. Patricia Larocca and Dr. Adrian César Razzitte from Universidad de Buenos Aires who have helped me in the task of reviewing and evaluating the chapters of the book, and both have been a great help.

**Dr. Ing. Walter Gustavo Fano**
Professor,
"Electromagnetic Radiation Laboratory",
Facultad de Ingenieria,
Electronic Department,
Universidad de Buenos Aires,
Argentina

**Adrian César Razzitte and Patricia Larocca**
University of Buenos Aires,
Argentina

Chapter 1

# Introductory Chapter: Causal Models of Electrical Permittivity and Magnetic Permeability

*Walter Gustavo Fano*

## 1. Fundamental concepts

The electricity and magnetism theory was formulated by a series of experimental physical laws, such as the Gauss's law of electrostatics, Ampere's law, Biot and Sabart's law, and Faraday's law, with the concepts of charges and electric currents that were used up to the middle of the 1800s. Since James C. Maxwell's *Treatise on Electricity and Magnetism*, with his contribution in the year 1873 [1], it was essential to formulate electromagnetic theory. This electromagnetic theory considers the addition of the displacement current to the conduction current to obtain the total current, which was a fundamental contribution to consider all the physical laws including the law of conservation of charge. Maxwell's equations are generally expressed differentially and are used considering the constitutive relationships, which are the relationships between the vectors of the electric and magnetic fields, which when applying the boarder conditions and the initial conditions, allow obtaining the solutions. These solutions are usually the electric and magnetic fields, since with these vector fields, the electric current, the electric potentials, the power, and other physical parameters of technological utility can be obtained. An issue that has been important in solving the cases that are found experimentally has been the electric and magnetic potentials, which allow the fields to be obtained many times in a simplified form. In electromagnetic theory, the so-called simple media are commonly used, whose characteristics are homogeneous, isotropic, and linear [2]. Here the properties of the media such as the electrical permittivity and the magnetic permeability of the constitutive relationships can be represented as complex numbers, where the electrical and magnetic losses are considered in the imaginary part. For cases of ferrous magnetic materials, for example, with losses, it is necessary to consider nonlinear behavior, although it will not be of interest in our study. Furthermore, the usual treatment of electromagnetic theory is done from the macroscopic point of view, although materials with electric or magnetic dipole moments are considered, because the treatment of quantum electromagnetism is already a specific topic.

## 2. Electromagnetic model of a material

Consider a material medium with an excitation of one electromagnetic wave, whose electric and magnetic fields vary over the time, it is considered that the input variables will be the electric or magnetic fields and the output variables will be the vectors of electric flux density and magnetic, respectively. The material can be considered as a system, with a specific transfer function, and this system is usually considered causal in physics, and from the point of view of the study of signals, it is called linear and time independence (LTI) [3]. These causal systems are important, because the Kramers-Kronig relations can be applied, which relate the real and imaginary part of the electrical permittivity and the magnetic permeability. The theoretical model of electrical permittivity and magnetic permeability of each media can be tested by mean of the Kramers-Kronig relations and Hilbert transform [3, 4]:

$$\varepsilon'(\omega) - 1 = \frac{1}{\pi} P \int_{-\infty}^{\infty} \frac{\varepsilon''(x)}{x - \omega} dx \tag{1}$$

$$\varepsilon''(\omega) = -\frac{1}{\pi} P \int_{-\infty}^{\infty} \frac{\varepsilon'(x) - 1}{x - \omega} dx \tag{2}$$

where P is the Cauchy principal value.

The fundamental assumption is known as the causality condition. The most primitive and intuitive one can be formulated as follows: the effect cannot precede the cause [5]. The numerical techniques now can allow the computation of Hilbert transform in order to test the electric permittivity model of the material.

## 3. Electromagnetic wave propagation

The interaction of electromagnetic waves with matter is an interesting topic to study several applications. Maxwell's equations allow to solve propagation problems in different media together with the boarder solutions that allow to obtain solutions in each application. In electromagnetic theory it is the development of the wave equation or D'Alembert's equation that is in the time domain, and as a function of frequency, we work with the Helmholtz equation, which, in the case of monochromatic sources, provides the two wave solutions, the wave vector and the propagation constant that allow to study the propagation in the different media, which are usually studied as perfect dielectrics or dielectrics without losses, perfect conductors, and dielectrics with losses. This last case of dielectric with losses is the one that has application to the topics of electromagnetic engineering, optoelectronic engineering, RF engineering, and communications engineering. The frequency of electromagnetic waves in technological applications is ranging from low-frequency waves, radio frequencies, microwaves, optical frequencies, infrared, ultraviolet, and even higher to high-energy frequencies. The energy associated with the electromagnetic wave is proportional to the propagation frequency using the Plank constant. The electromagnetic waves that affect an interface from the air to the dielectric medium that usually has losses will be reflected energy and transmitted to the medium under study, dissipating heat in the medium, and it will attenuate the amplitude of the electromagnetic wave that propagates and causes the dispersion effect. This means that the propagating signal will have different propagation speeds for different frequencies, causing a distortion of the propagating signal as it moves through the dispersive medium. Knowledge of the electrical and magnetic properties, which are intrinsic properties of matter, such as the response of materials to be used in the electronic industry, are essential for the design and construction of electrical and electronic devices. The materials in the transmission lines, waveguides, and fiber optics where an electromagnetic wave propagates in the infrared band have materials that have losses and dispersion that must be considered. In the case of an alternating current flowing in the soil, it will also be necessary to consider the electrical properties of the soil as the electrical conductivity for the various applications in electrical engineering. The application of heat to a junction of two metals or two semiconductors produces a potential difference at the ends; this phenomenon is called the Seebeck effect, which in metals the potential differences obtained are very small. For this reason, new composite materials that can obtain a higher Seebeck coefficient are investigated. In metals the Seebeck coefficient is generally of the order of $1\mu V/C$; it increases greatly in cases where a metal is measured with a semiconductor, for example. Currently, a technological application of this effect is thermocouples, which are used to measure temperature.

## 4. Organization of the book

In chapter I of the book, the physical sense of the phenomenon of dispersion of electromagnetic waves is discussed; the group speed is obtained. Then from the Lorentz force, the plasma model and the dispersion in the plasma, and in a conductive medium, are discussed. Dispersion topics that are of greatest technological interest are discussed, such as modal, chromatic, and intramodal dispersion. Chapter II studies electromagnetic propagation through the soil, where historically it was used for telegraphic transmissions, in the transmission of surface waves in the AM bands; the knowledge of the electrical properties of soil are applied to the study of agriculture and archeology, which have become very relevant these days. In this chapter the different methods of measuring the electrical properties of the soil are discussed. A widely used technique is time domain reflectometry, which studies the response of the reflected pulses in the time domain to obtain the electrical properties of the soil. Another way to obtain the electrical properties of the soil is by measuring the impedance in the frequency domain of a transmission line known and built for this purpose. In this chapter the own experimental results obtained by the author are presented. In chapter III the electrical conductivity in direct current in molten salts ("Electrical Conductivity of Molten Salts") is studied from a microscopic point of view using the Langevin equation, which implies a time-dependent memory function $\gamma(t)$ in relation to the friction forces acting on the constituent ions under the electric field. The properties of the ionic liquid transport phenomenon are important for industry and technological applications. Ionic liquids are divided into two main groups: molten salts and electrolytic solutions. Chapter IV deals with the interaction of electromagnetic waves with the biological tissues of human beings and the skin of animals. Electromagnetic waves can come from the sun, and frequencies range from very low frequencies to gamma ray frequencies. As it is well-known, the atmosphere filters the highest energy frequencies such as gamma rays, X rays, or ultraviolet rays. This work deals with and studies the transmission, reflection, and reflection coefficients in the skin of humans and animals of electromagnetic waves. In chapter V, we work with the Seebeck effect, which is about two metals or semiconductors to which different temperatures are applied, and a potential difference is produced. The reverse effect is called Peltier and consists of applying a potential difference to the conductors/semiconductors, and heating or cooling occurs at the junction. These thermoelectric properties have technological applications that are being used such as thermocouple temperature sensing and Peltier effect cooling systems. Here we present new thermoelectric materials tested as tellurium telluride chalcogenide nano-materials.

## Author details

Walter Gustavo Fano
Electromagnetic Radiation Laboratory, Facultad de Ingeniería, Universidad de Buenos Aires, Buenos Aires, Argentina

*Address all correspondence to: gfano@fi.uba.ar

## References

[1] Stratton JA. Electromagnetic Theory. Hoboken, New Jersey, USA: McGraw-Hill Book Company; 2007

[2] Trainotti V, Fano WG. Ingenieria Electromagnetica. 1st ed. Vol. 1. Buenos Aires, Argentina: Nueva Libreria; 2004

[3] Fano WG, Boggi S, Razzitte AC. Causality study and numerical response of the magnetic permeability as a function of the frequency of ferrites using Kramers-Kronig relations. Physica B. 2008;**403**:526-530

[4] Landau LD, Lifchitz EM. Electrodynamics of Continuous Media. Boston, USA: Addison Wesley; 1981

[5] Nussenzveig HM. Causality and Dispersion Relations. New York: Academic Press; 1972

Chapter 2

# Fields in Dispersive Media

*Emeka Ikpeazu*


## Abstract

It is not just the case that matter affects the propagation of light—or more specifically electromagnetic (EM) radiation—it is also the case that light affects the matter through which it propagates. Conversely, this affects the propagation of light through the medium, but in a much more specific way; this effect is a function of the properties of both the material and the incident EM radiation. We will additionally discuss the effects of dispersion in confined (bounded) media, i.e., where the dispersion is a function of the arrangement of certain materials and unbounded media where EM radiation is free to propagate undisturbed. This will be important when we discuss the propagation electric field signals of such media as well.


**Keywords:** refraction, materials, dispersion, electromagnetics, wave propagation, fiber optics

## 1. Introduction

Incident electromagnetic (EM) radiation excites the molecules in a material, and these molecules become polarized; they respond according to the direction of the electric field. **Figure 1** shows the initial step of the process of material polarization.

It should be adequately noted that there are *several* ways that material response to light-matter interaction can be analyzed. For the purposes of relevance, this chapter will analyze the mechanical response of the atom—as illustrated in **Figure 2**—in response to incident EM radiation waves.

In this chapter we will discuss polarization at the level of the interacting material, the effects of dispersion on a train of pulses, various types of polarization, and methods for reducing polarization to maintain the integrity of optical signals. But first, some words on non-dispersion in unbounded media, a more general term for what would normally be called "free space."

Unbounded media are the baseline of understanding EM wave propagation. In unbounded media, waves are free to propagate unperturbed. Examples of unbounded media include the ocean, the air, and outer space. An unbounded medium would be the ideal location for an isotropic antenna as the radiation would be free to propagate in all directions, only weakening in accordance with the inverse square law. Such is ideal for radio towers which produce low-frequency EM waves which can propagate for kilometers and reach many people.

Additionally, unbounded media are generally non-dispersive. This is to say that the speed of energy propagation is orthogonal to the frequency of the said propagating energy. In the previous paragraph, it was said the waves in unbounded media

**Figure 1.**
*An EM traveling wave is incident on a molecule of a particular material. This induces a unique response in molecules of the material.*

**Figure 2.**
*The mechanical response of the material to incoming monochromatic EM radiation. The vibrational response of the molecule is a function of its properties. This same principle is also related to the color of the flame that a molecule produces [1].*

are free to propagate unperturbed. This is not *completely* true. Natural disturbances *within* the air or sea can interfere with the propagation of energy therein. However, these natural impediments are not necessarily treated as features of the media itself. It is for this reason that direct or *line-of-sight* propagation is key in facilitating communication between antennas and cell towers.

## 2. Understanding dispersion

Dispersion can be difficult to understand. There are picture balls on two parallel infinite conveyor belts running at different speeds, Belt A at the top and Belt B at the bottom. The conveyor belts in **Figure 3** carry balls at 2.5 cm/s and 3.0 cm/s, respectively. Additionally, the centers of the balls on each belt are separated by 10 cm and 12.5 cm, respectively. At time $t = 0$, the centers of Red Ball 0 ($RB_0$) and Blue Ball 0 ($BB_0$) are aligned. However, the center of $BB_{-1}$ trails that of $RB_{-1}$ by a distance of 2.5 cm.

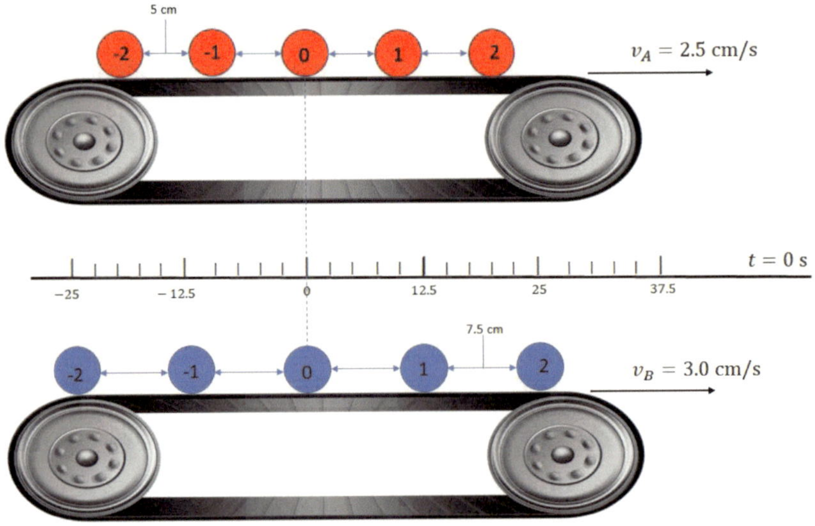


**Figure 3.**
*A mechanical kinematic illustration of the principle of the dispersion.*

Since Belt B is traveling faster than Belt A, the distance between $BB_{-1}$ and $RB_{-1}$ will decrease. From the perspective of Belt A, Belt B is moving with a velocity of 0.5 cm/s. At time $t = 5$, the center of $BB_{-1}$ is aligned with that of $RB_{-1}$; however, in that time the balls on Belt B have moved a net 15 cm, while those on Belt A have moved a net 12.5 cm. However, if one were to see the conveyor belts as a group, the belts would appear to have moved 2.5 cm (measuring from location of alignment) within that time, giving the conveyor belts a "group velocity" of 0.5 m/s even though the individual components are moving faster.

The situation described is analogous to the behavior of a wave propagating in free space. As the individual waves travel with their respective phase velocities along the guide, the two peaks become disaligned for a period [2]. Because the center of $RB_{-1}$ will have traveled a distance $\lambda_{RB} + \Delta z$ with a phase velocity $v_{zRB}$ and the center of $BB_{-1}$ will have traveled a distance $\lambda_{BB} + \Delta z$ with phase velocity $v_{zBB}$, one can then write a set of equations as

$$\lambda_{RB} + \Delta z = v_{z,RB}\Delta t \tag{1}$$

$$\lambda_{BB} + \Delta z = v_{z,BB}\Delta t \tag{2}$$

We can then write $\Delta t$ as

$$\Delta t = \frac{\lambda_{RB} - \lambda_{BB}}{v_{z,RB} - v_{z,BB}} \tag{3}$$

In the same way, one can express $\Delta z$ as

$$\Delta z = \frac{\lambda_{RB} v_{z,BB} - \lambda_{BB} v_{z,RB}}{v_{z,RB} - v_{z,BB}} \tag{4}$$

Finally, one can express the group velocity as

$$\frac{\Delta z}{\Delta t} = v_z = \frac{\lambda_{RB} v_{z,BB} - \lambda_{BB} v_{z,RB}}{\lambda_{RB} - \lambda_{BB}} = \frac{\lambda_{RB}\lambda_{BB} f_{BB} - \lambda_{BB}\lambda_{RB} f_{RB}}{\lambda_{RB} - \lambda_{BB}} \tag{5}$$
$$= \frac{\lambda_{BB}\lambda_{RB}\left(f_{BB} - f_{RB}\right)}{\lambda_{RB} - \lambda_{BB}} = \frac{f_{BB} - f_{RB}}{\frac{1}{\lambda_{BB}} - \frac{1}{\lambda_{RB}}} \to \frac{\omega_{BB} - \omega_{RB}}{2\pi\left(\frac{1}{\lambda_{BB}} - \frac{1}{\lambda_{RB}}\right)}$$

**Figure 4.**
*The separation of white light into different colors by prismatic dispersion.*

We know that the wavevector $k = 2\pi/\lambda$, so one express the group velocity as

$$v_g = \frac{d\omega}{dk} \tag{6}$$

This is to say the group velocity is the degree change in the temporal frequency for every change in the spatial frequency. What this means is that the value of the frequency is on some level a function of the $k$-vector and ultimately a function of the wavelength. As illustrated in **Figure 4**, this has implications for the propagation of light through various media as different wavelengths will travel at different velocities and take various pathways therein [3].

## 3. Plasmonic dispersion

The Lorentz force is the force on a point charge due to EM fields. As discussed in the Introduction, EM radiation causes motion among the particles in a material as shown in **Figure 2**. The direction (vector) of this force is called the polarization. Assuming a sea of free particles in vacuum [4], the strength of the Lorentz force on a point charge is expressed as

$$\vec{\boldsymbol{F}} = q\left(\vec{\boldsymbol{E}} + \vec{\boldsymbol{v}} \times \vec{\boldsymbol{B}}\right), \tag{7}$$

where $\vec{\boldsymbol{F}}$, $q$, $\vec{\mathbf{E}}$, $\vec{\boldsymbol{v}}$, and $\vec{\mathbf{B}}$ are the force, the charge, the electric field strength, the particle velocity, and the magnetic flux density, respectively. For our purposes, we consider an isotropic material where the electric permittivity is a simple scalar $\epsilon$. This means that the Lorentz force vector will be unidirectional. In birefringent and anisotropic, the permittivity would be expressed as a tensor, and the Lorentz force vector would be multidirectional.

We additionally assume a dielectric unmagnetized material such that the value of $\vec{\mathbf{B}}$ is assumed to be zero. This makes the response of the atoms into a simple harmonic oscillator. In this case we can express the Lorentz force as simply

$$\vec{\boldsymbol{F}} = q\vec{\boldsymbol{E}} = m\vec{a} = m\left(\frac{d^2\vec{\boldsymbol{d}}}{dt^2}\right) = m(j\omega)^2\vec{\boldsymbol{d}} = -m\omega^2\vec{\boldsymbol{d}}, \tag{8}$$

where $m$, $\omega$, and $\vec{\boldsymbol{d}}$ are the mass, the angular frequency, and dipole vector, respectively. We can simplify to get the dipole vector as

$$\vec{\boldsymbol{d}} = -\frac{q\vec{\mathbf{E}}}{m\omega^2} \rightarrow \vec{\mathbf{P}} = Nq\vec{\boldsymbol{d}} = -\frac{Nq^2\vec{\mathbf{E}}}{m\omega^2}, \tag{9}$$

where $\vec{\mathbf{P}}$ is the polarization vector and $N$ is the dipole density. The electric displacement field $\vec{\mathbf{D}}$ is the permittivity $\epsilon$ multiplied by the electric field $\vec{\mathbf{E}}$, but in dielectric materials, the electric field induces a response in the material known as the polarization. This effect is added to the original displacement field to make it so that

$$\vec{\boldsymbol{D}} = \epsilon_0\vec{\boldsymbol{E}} + \vec{\boldsymbol{P}} = \left(\epsilon_0 - \frac{Nq^2}{m\omega^2}\right)\vec{\boldsymbol{E}} \tag{10}$$

As a result, the *effective* permittivity is expressed in (11) as

$$\epsilon = \epsilon_0 \left(1 - \frac{Nq^2}{\epsilon_0 m \omega^2}\right) = \epsilon_0 \left(1 - \frac{\omega_p^2}{\omega^2}\right), \tag{11}$$

where the plasma frequency $\omega_p = q\sqrt{N/\epsilon_0 m}$.

The plasma frequency is the natural oscillation frequency of the displaced electrons in a neutral plasma [4, 5] of free electrons where it is assumed that collisions are rare. Altogether we can express the dispersion relation thusly

$$k^2 = \omega^2 \epsilon \mu = \epsilon_0 \mu_0 \left(\omega^2 - \omega_p^2\right) \rightarrow \omega(k) = \sqrt{k^2 c^2 + \omega_p^2} \tag{12}$$

This means that the group velocity is expressed as

$$v_g = \frac{kc^2}{\sqrt{k^2 c^2 + \omega_p^2}} = c\sqrt{1 - \left(\frac{\omega_p^2}{\omega^2}\right)} \tag{13}$$

In a material like germanium (Ge) the intrinsic carrier density $N_{\mathrm{Ge}}$ is $2.5 \times 10^{19} \mathrm{m}^{-3}$ [6]; the rest of the variables are known constants. This gives germanium a plasma frequency of $f_{p,\mathrm{Ge}} = 44.8$ GHz. We apply the relation in (13) to germanium to measure the frequency-dependent group velocity of light therein as demonstrated in **Figure 5**.

Below the plasma frequency, the value of the $k$-vector becomes purely imaginary, meaning that the electric field is purely evanescent—and non-propagating—at that point. The group velocity is the slope of the dispersion curve in **Figure 6**. For both curves, the lowest allowable frequency is the plasma frequency. Below this frequency, the permittivity is negative, the $k$-vector is imaginary, and,

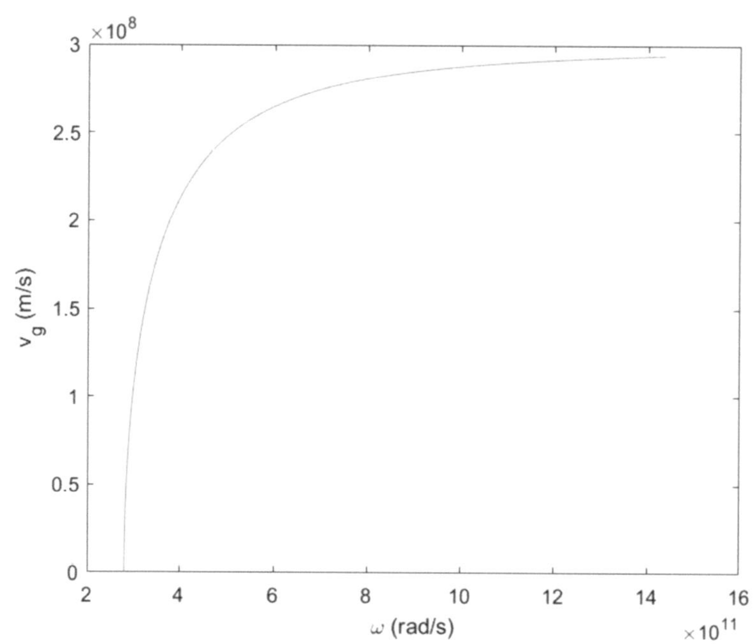


**Figure 5.**
*The group velocity of light through germanium as a function of frequency.*

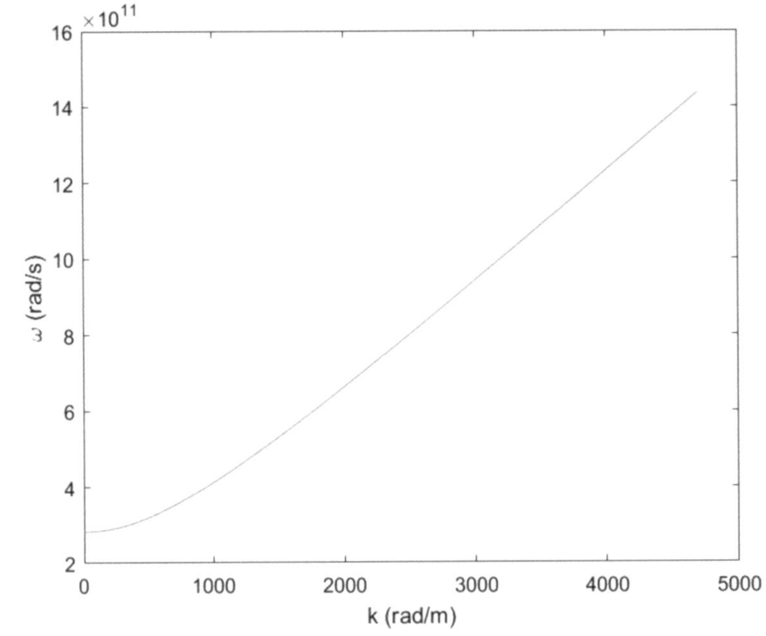


**Figure 6.**
*The dispersion relation for germanium where ω is function of k.*

consequently, the field simply evanesces into the material and propagation completely ceases.

## 4. Dispersion in conductive media

In the previous section, we derived plasmonic dispersion from understanding the Lorentz force law as it pertains to electrons in a dielectric material. Dispersion can also occur in a conducting material where the charges therein are unbound. When we apply our knowledge of electromagnetics and electrostatics to a conductor, we discover a few things:

1. There is zero net charge within a conductor, although charges can still randomly move around therein as seen in **Figure 7**.

2. All the charge rests on the surface of the conductor, perpendicular to the surface.

3. The presence of an electric field induces the motion of electrons, creating a conduction current [7].

Additionally, in a conductor the free charges are always pushed to the surface so that electrostatic equilibrium is maintained. This means that the gradient of the electric field $\vec{\nabla} \cdot \vec{\mathbf{E}}$ equals zero. Things are different in the presence of an electric field, e.g., **Figure 8**, where the electrons are pulled in the opposite direction of the electric field.

With regard to conductive media, we can do the same thing through an understanding and application of Faraday's law which states that a time-varying magnetic flux density *always* accompanies a nonconservative electric field, i.e., a field $\vec{\mathbf{E}}$ for which $\vec{\nabla} \times \vec{\mathbf{E}} = \mathbf{0}$. This is demonstrated by the equation

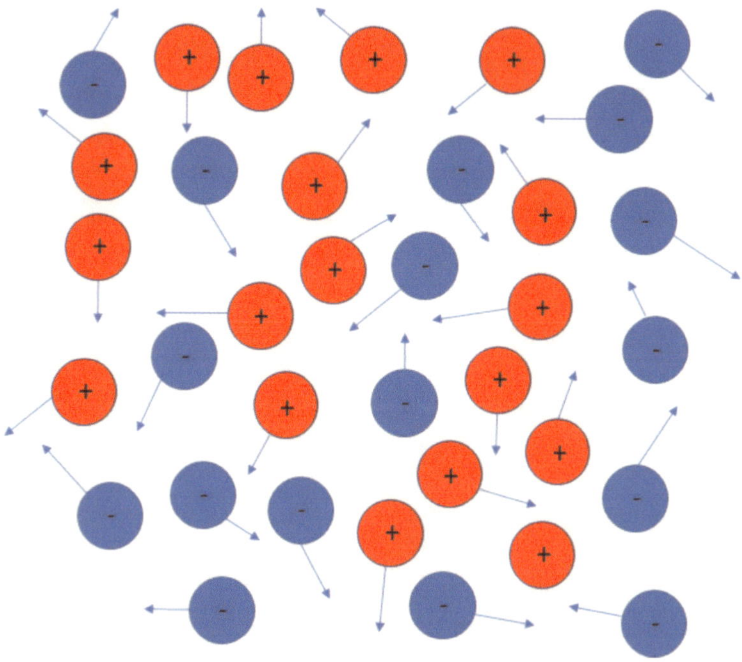

**Figure 7.**
*The distribution of freely moving protons (red) and electrons (blue) in a conductive material in electrostatic equilibrium where* $\mathbf{Q}_{\boldsymbol{\Sigma}} = 0$.

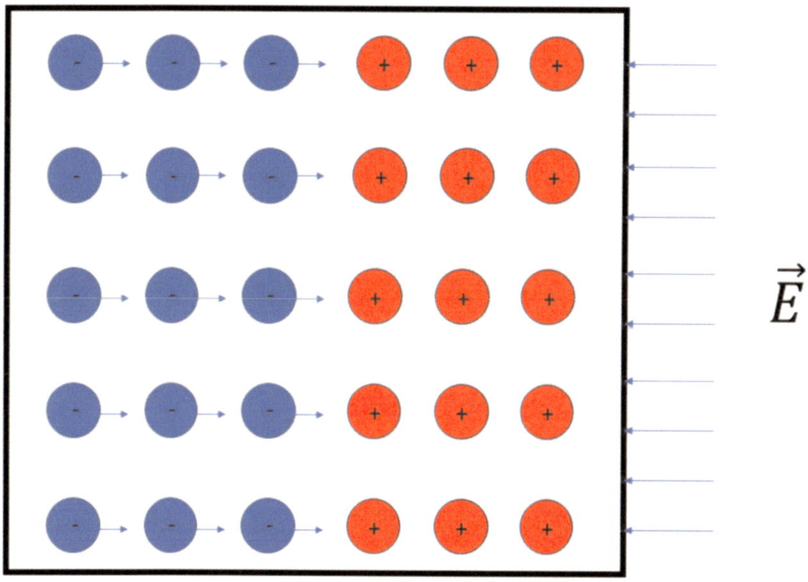


**Figure 8.**
*The kinetic behavior of electrons in the presence of an electric field produces a steady-state current known as the conduction current.*

$$\vec{\nabla} \times \vec{\boldsymbol{E}} = -\frac{\partial \vec{\boldsymbol{B}}}{\partial t} = -\mu_0 \frac{\partial \vec{\boldsymbol{H}}}{\partial t} \tag{14}$$

We can combine this with Ampère's law where the curl of the magnetic field generates a time-varying electric field such that

$$\vec{\nabla} \times \vec{H} = \vec{J}_f + \epsilon \frac{\partial \vec{E}}{\partial t} = \sigma \vec{E} + \frac{\partial \vec{E}}{\partial t} = (\sigma + j\omega\epsilon)\vec{E} = j\omega\epsilon_{eff}\vec{E}, \tag{15}$$

where $\vec{J}_f$ is the free current density vector produced in the presence of an electric field as exemplified in **Figure 8** and $\sigma$ is the conductivity. We can calculate the effective permittivity this way such that $\epsilon_{eff} = \epsilon\left(1 - \frac{j\sigma}{\omega\epsilon}\right)$. In this way we can write the dispersion relation.

$$\begin{aligned} k &= \omega\sqrt{\mu_0\epsilon_{ff}(\omega)} = \omega\sqrt{\frac{\mu_0\epsilon}{\epsilon_0}}\sqrt{1 - \frac{j\sigma}{\omega\epsilon}} = \sqrt{\frac{\mu_0\epsilon}{\epsilon_0}}\sqrt{\omega^2 - \frac{j\sigma\omega}{\epsilon}} \\ &= \sqrt{\frac{\mu_0\epsilon}{\epsilon_0}}\sqrt[4]{\omega^4 + \left(\frac{\sigma^2\omega^2}{\epsilon^2}\right)}e^{-i\left(\frac{tan^{-1}\left(\frac{\sigma}{\epsilon\omega}\right)}{2}\right)} \end{aligned} \tag{16}$$

This wavenumber $k$ is also complex where $k = k' - jk''$, where

$$\begin{aligned} k' &= \sqrt{\frac{\mu_0\epsilon}{\epsilon_0}}\sqrt[4]{\omega^4 + \left(\frac{\sigma^2\omega^2}{\epsilon^2}\right)}\cos\left(\frac{\tan^{-1}\frac{\sigma}{\epsilon\omega}}{2}\right) \leftrightarrow \\ k'' &= \sqrt{\frac{\mu_0\epsilon}{\epsilon_0}}\sqrt[4]{\omega^4 + \left(\frac{\sigma^2\omega^2}{\epsilon^2}\right)}\sin\left(\frac{\tan^{-1}\frac{\sigma}{\epsilon\omega}}{2}\right). \end{aligned} \tag{17}$$

The real part of the wavenumber $k'$ represents wavenumber inside the material and the imaginary part $k''$ represents the attenuation coefficient inside the material.

$$\begin{aligned} v_g &= \left(\frac{dk'}{d\omega}\right)^{-1} \\ &= \sqrt{\frac{\epsilon_0}{\mu_0\epsilon}}\left\{\frac{\left(\frac{\sigma^2}{\epsilon^2} + 2\omega^2\right)\cos\left(\frac{\tan^{-1}\frac{\sigma}{\epsilon\omega}}{2}\right)}{2\omega\left(\frac{\sigma^2}{\epsilon_0^2} + \omega^2\right)} + \frac{\frac{\sigma}{\epsilon}\sqrt[4]{\omega^4 + \left(\frac{\sigma^2\omega^2}{\epsilon^2}\right)}}{\left(\omega^2 + \frac{\sigma^2}{\epsilon^2}\right)}\sin\left(\frac{\tan^{-1}\frac{\sigma}{\epsilon\omega}}{2}\right)\right\}^{-1} \\ &= \frac{2\sqrt{\frac{\epsilon_0}{\mu_0\epsilon}}\left(\omega^2 + \frac{\sigma^2}{\epsilon^2}\right)}{\left(\frac{\sigma^2}{\epsilon^2\omega} + 2\omega\right)cos\left(\frac{tan^{-1}\frac{\sigma}{\epsilon\omega}}{2}\right) + \frac{\sigma}{\epsilon}\sqrt[4]{\omega^4 + \left(\frac{\sigma^2\omega^2}{\epsilon^2}\right)}sin\left(\frac{tan^{-1}\frac{\sigma}{\epsilon\omega}}{2}\right)} \end{aligned} \tag{18}$$

Of course, there are various conditions that could simplify Eq. (16). The $\sigma/\omega\epsilon$ component in (15) is known as the dielectric *loss tangent* $\tan\delta$ [8]. This figure $\delta$ describes a materials inherent ability to dissipate electromagnetic energy within it. For example, a material with a high conductivity will absorb—dissipate—EM radiation more quickly. Such a material has would have a high loss tangent compared to a good dielectric.

We can express the dielectric relaxation time $\tau_d$ as $\epsilon/\sigma$. In cases that $\delta \ll 1$ or $\tau_d^{-1} \ll \omega$—as would be the case in a lossy dielectric—we can approximate the permittivity as $\epsilon_{eff} \approx 1 - \frac{j\sigma}{2\omega\epsilon}$ using the binomial expansion theorem. The material behaves like a standard dielectric as $v_g = c/n$, where $n$ is the index of refraction, and $k'' = \sigma\eta/2$, where $\eta$ is the impedance of the material $\sqrt{\mu_0/\epsilon}$.

In cases where $1 \ll \delta$ or $\omega \ll \tau_d^{-1}$, the case of a good conductor, $k \approx \sqrt{\sigma\mu_0\omega/2}(1 - j)$ and $v_g \approx \sqrt{8\omega/\sigma\mu_0}$. In the latter case, the field penetrates only slightly into the material as $\mathcal{O}(k') = \mathcal{O}(k'')$.

$$\frac{2\sqrt{\frac{\epsilon_0}{\mu_0\epsilon}}\left(\omega^2+\frac{\sigma^2}{\epsilon^2}\right)}{\left(\frac{\sigma^2}{\epsilon^2\omega}+2\omega\right)cos\left(\frac{tan^{-1}\frac{\sigma}{\epsilon\omega}}{2}\right)+\frac{\sigma}{\epsilon}\sqrt[4]{\omega^4+\left(\frac{\sigma^2\omega^2}{\epsilon^2}\right)}sin\left(\frac{tan^{-1}\frac{\sigma}{\epsilon\omega}}{2}\right)} \tag{19}$$

Additionally, the attenuation is a new phenomenon resulting from the fact that the $k$-vector is complex. When we take the reciprocal of this coefficient get another quantity known as the *skin depth* or penetration depth $\delta_p$, the extent to which the field penetrates the material. In a good conductor, $\delta_p$ takes a value $\sqrt{2/\sigma\mu_0\omega}$. The values of $\epsilon$ and $\sigma$ are listed in **Table 1** for some good conductors [9].

Good conductors are characterized by an abundance of free electrons. Good conductors (good dielectrics), on the other hand, are characterized by a paucity of such electrons. Examples of good dielectrics are enumerated in **Table 2** [10].

**Figure 9** shows the interaction between an electric field and an good conductor (e.g., gold) at a wavelength of 1.55μm. The field decays very rapidly as gold has a skin depth of 5.66 nm and a dielectric relaxation time of 1.49 attoseconds ($10^{-18}$ s) at this wavelength. This contrasts with a lossy dielectric where the skin depth is much larger, and there is thus more field penetration as displayed in **Figure 10** with germanium where the skin depth is roughly a centimeter and the dielectric relaxation time is 66.1 picoseconds.

Conductive media have more free electrons that *move* in the presence of an electric field than those that simply vibrate. Conductive media produce a loss (attenuation) component in propagation of incoming EM radiation. Additionally, the drift movement of electrons in conductive media causes heat to which the electrical energy from the incoming field is transferred. This energy conversion causes the dissipation of the field. Nevertheless, in the context of the phenomenon of dispersion, the degree of this attenuation is a function of the frequency of the incoming EM radiation.

| Material | Conductivity (S/m) @ $T = 293$ K |
|---|---|
| Silver | $6.1 \times 10^7$ |
| Copper | $5.8 \times 10^7$ |
| Gold | $4.1 \times 10^7$ |
| Aluminum | $3.5 \times 10^7$ |
| Brass | $1.5 \times 10^7$ |

**Table 1.**
*Conductivities of some good conductors.*

| Material | Permittivity (F/m) |
|---|---|
| Calcium | $2.685 \times 10^{-11}$ |
| Marble | $7.08 \times 10^{-11}$ |
| Silicon dioxide ($SiO_2$) | $3.452 \times 10^{-11}$ |
| Slate | $3.54 \times 10^{-11}$ |
| Polytetrafluoroethylene (PTFE) | $1.77 \times 10^{-11}$ |

**Table 2.**
*Conductivities of some good dielectrics.*

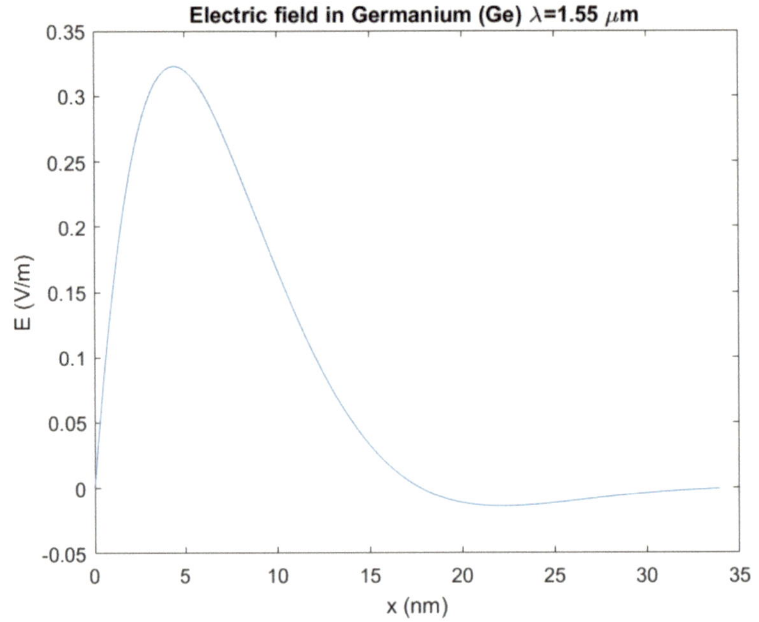


**Figure 9.**
*The rapid decay of an electric field upon impact interaction with gold.*

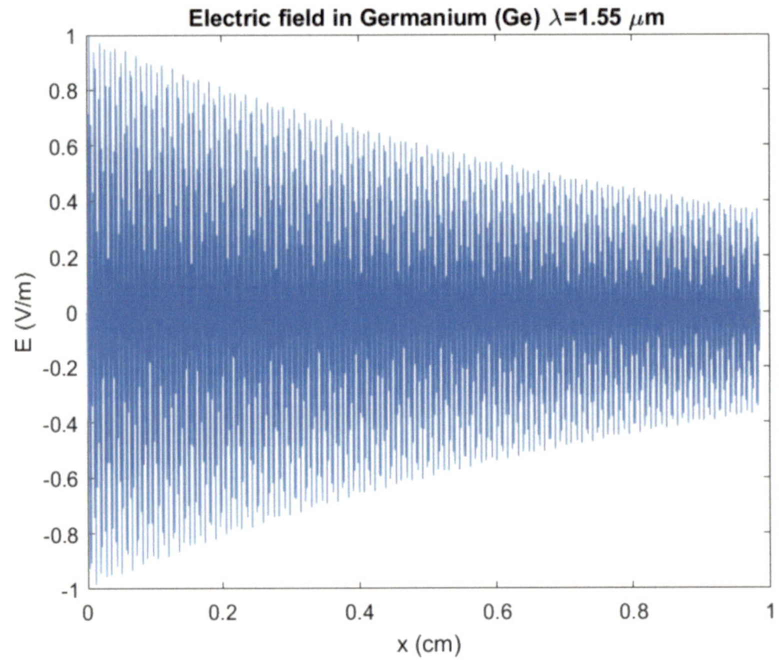


**Figure 10.**
*The comparatively slow decay of an electric field upon impact interaction with germanium.*

## 5. Modal dispersion

So far, this chapter has covered the form that dispersion takes in media (i.e., material dispersion). However, there are other types of dispersion that exist which ultimately stem from the properties of the material. One of these other ways is modal dispersion [11]. This is when the velocity of various modes changes as a function of the input wavelength. As displayed in **Figure 11**, the different modes that propagate at different group velocities and angles, which themselves are functions of an input wavelength. These modes that arise in a waveguide are the different forms that the electric field can take in a geometrically confined structure. These forms are functions of satisfying the Helmholtz equation for the geometry of the confining structure. Only a finite number of fields do this and produce the appropriate wavevector in the $z$-direction, $\boldsymbol{\beta}$. Consequently, only a finite number of modes will be allowed in the structure.

The angle at which certain modes propagate within makes a difference as to the speed of propagation because shallower angles have less deviation from the center equilibrium as shown in **Figure 12**.

In a rectangular waveguide, the total distance a mode must travel is expressed as

$$L = \frac{L}{\cos\theta_x \cos\theta_y} \tag{20}$$

where $\boldsymbol{\theta_x}$ and $\boldsymbol{\theta_y}$ are the angles of deviation in the $x$- and $y$-directions, respectively. The higher the order of the mode, the steeper the angle of the travel, and the more distance the mode must traverse through the waveguide. In **Figure 13**, the same principle holds true in an optical fiber. The rays of the fundamental modes (TM and TE) take a meridional form (a) passing through origin as they propagate and the hybrid modes—HE and EH—spiral in a helical form taking a twisted path through the fiber.

Additionally, modes that have a shallower angle of interaction with the core-cladding boundary will have higher group velocities. A higher-order mode has a

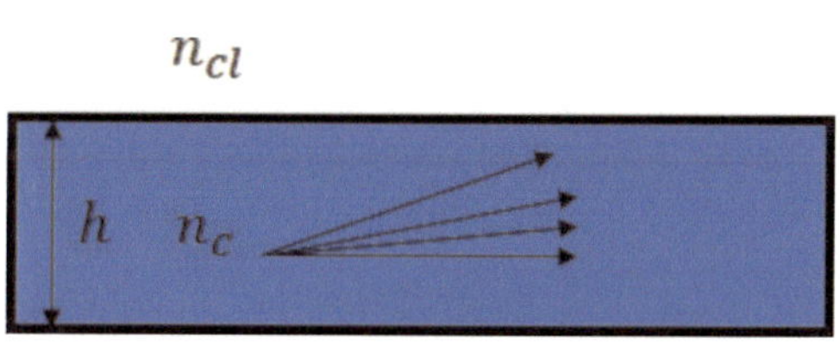


**Figure 11.**
*The various modes within a step-index waveguide—where h, $\boldsymbol{n_c}$, and $\boldsymbol{n_{cl}}$ are the height and the core and cladding indices, respectively.*

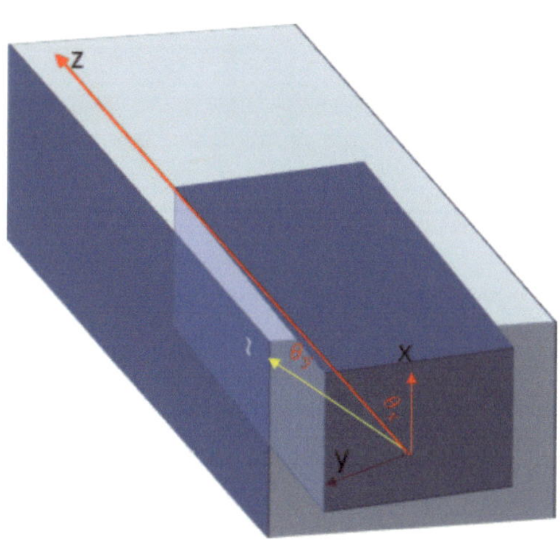


**Figure 12.**
*A mode in a rectangular waveguide propagates in the x- and y-directions at various angles.*

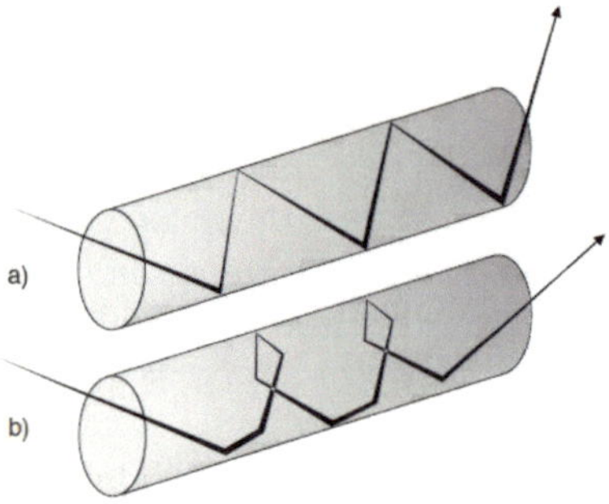


**Figure 13.**
*Rays taking meridional (a) and helical (b) forms in an optical fiber [12].*

lower $\beta$ value as its $\boldsymbol{k}$-vectors in other directions are higher, leading to steeper interaction angles with the core-cladding boundary.

The way that the modes propagate through the fiber at different velocities makes the waveguide an effective multipath propagation channel where the received signal is a sum of the various scaled "echoes" of the original signal carried by the various modes that propagate therein.

As a result, this dispersion distorts the pulse and corrupts the information. As shown in **Figures 14** and **15**, the pulses deform from their respective Gaussian and rectangular forms as a result of the different path delays, a phenomenon known as *delay distortion* [13]. Calculation of modal delays is used in order to set various bit rates for optical pulses.

One way to ameliorate modal (intermodal) dispersion is to use a single-mode waveguide where the only one mode is carried through. Another way is to make use of waveguides with a parabolic index of refraction, producing a quasi-sinusoidal pathway through the waveguide as demonstrated in **Figure 16**.

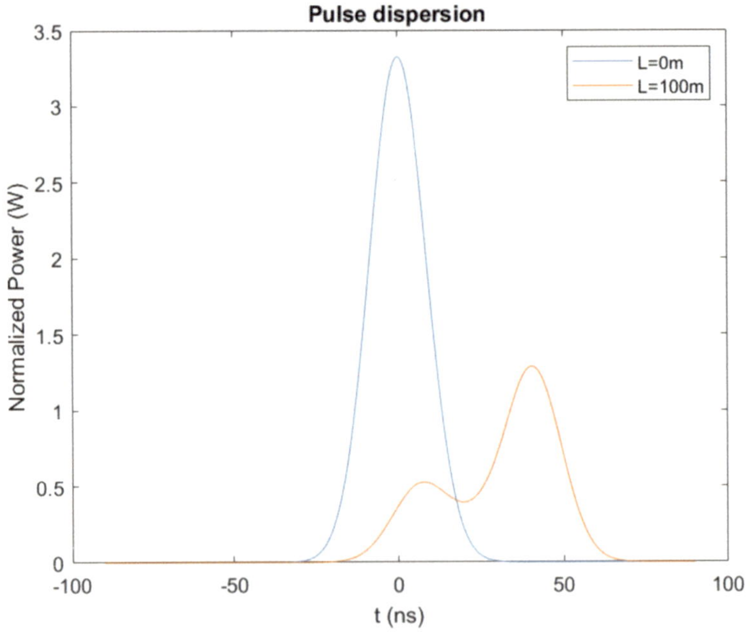


**Figure 14.**
*The intermodally dispersive time evolution of a Gaussian optical pulse in a multimode rectangular waveguide.*

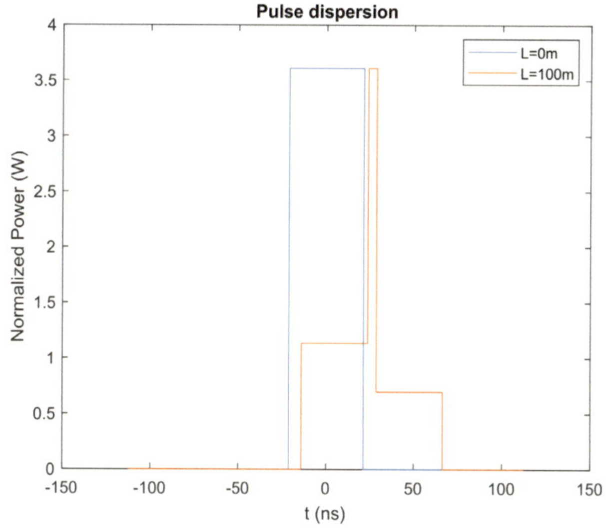


**Figure 15.**
*The intermodally dispersive time evolution of the more commonly used rectangular pulse in a multimode rectangular waveguide.*

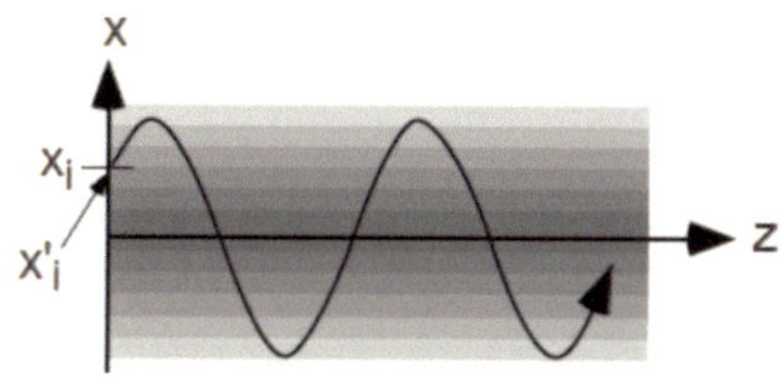


**Figure 16.**
*The pathway through the waveguide is sinusoidal [14] as it is bound to its axis.*

The ray switches paths as it passes the region of highest index, meaning that it spends less of its time in the region of lowest index. This is to say that the phase velocity of the wave changes within the waveguide. From a ray optics perspective, the crest of the wave has a position-varying velocity, effectively making continuous switches through "different materials."

The dispersion characteristic for graded-index (GRIN) waveguides is quite different as well. **Figure 17** shows that higher wavelengths are correlated with less dispersion for GRIN waveguides. The step-index, on the other hand, maintains a positive wavelength dispersion characteristic.

It should be noted that modal dispersion is more of a function of the *arrangement* of media, not the sole medium itself. Changes in size, cladding material, and wavelength can affect the propagation of light therein.

## 6. Chromatic dispersion

Changes in wavelength are important in understanding how light will propagate. As shown in **Figure 18**, the way these $\beta$ values vary with the wavelength produces various group velocities for different modes. The higher-order modes have a lower group velocity as they travel. The index of the fused silica core in the waveguide

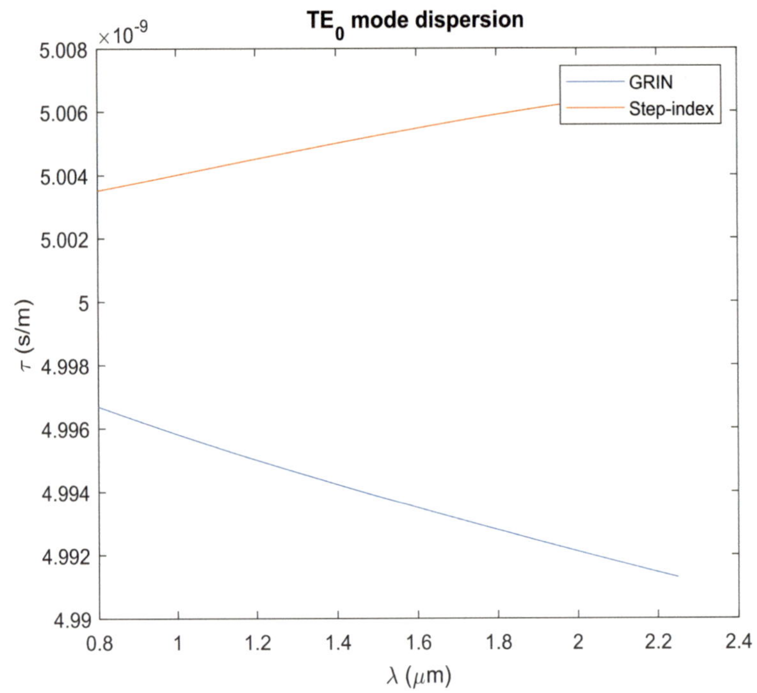


**Figure 17.**
*The fundamental mode ($TE_0$) dispersion characteristic of GRIN slab waveguide vs. that of a step-index slab waveguide.*

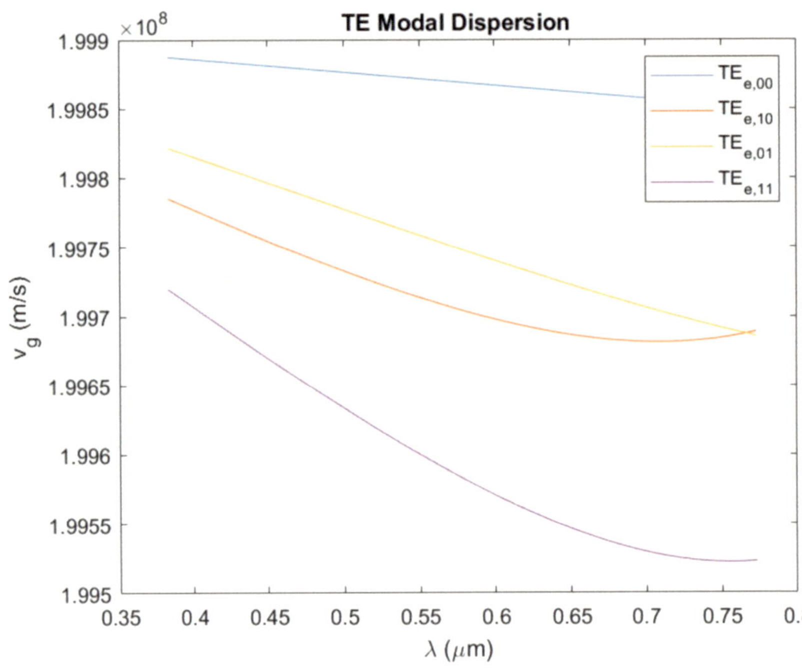


**Figure 18.**
*The group velocities of the transverse electric (TE) modes in a fused silica rectangular waveguide as a function of the wavelengths in the visible spectrum.*

described in **Figure 18** is 1.5, meaning that light will travel roughly $\mathbf{2 \times 10^8}$ m/s therein. However, the contribution from dispersion increases the *effective* index $\boldsymbol{n_{eff}}$ "experienced" by each mode at a particular wavelength.

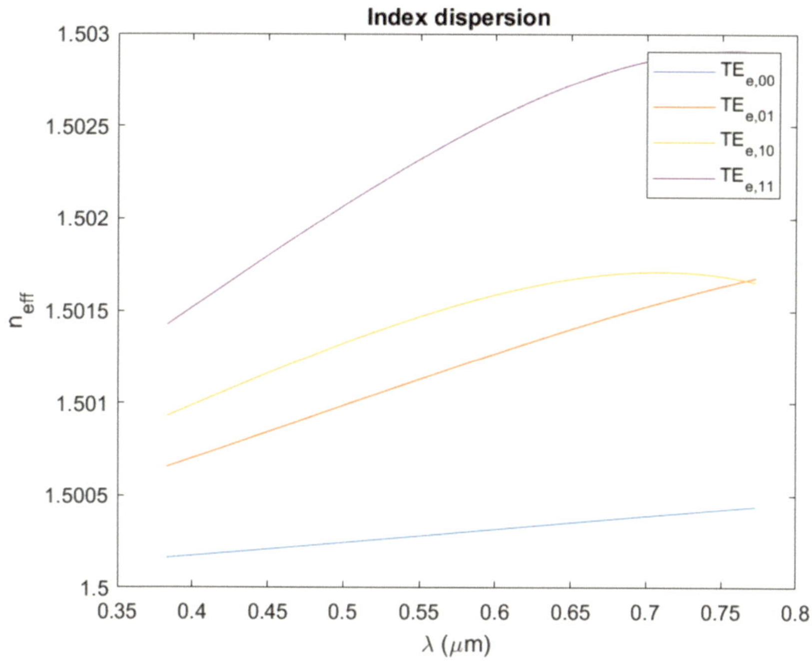


**Figure 19.**
*Dispersion creates a higher refractive index for the modes as λ increases.*

The plot in **Figure 18** can be reconfigured to show that each mode has an effective index of refraction in the core that varies with its wavelength. This means that different "colors" of light will have different refractive indices in the waveguide. **Figure 19** shows that as $\lambda$ increases, the group velocity decreases.

**Figure 19** shows that there is an added index of refraction that comes from the dispersion that contributes the delay of the modes; this is also known as the group index $N_g$.

## 7. Intramodal dispersion

Dispersion can occur even in the context of a single-mode waveguide (of whatever geometry). In the previous section, we demonstrated a difference in $n_{eff}$ as it pertains to various wavelengths stimulating a particular mode. However, we can observe a similar phenomenon in a single-mode fiber.

In the Introduction, we discussed how one can model the plasmonic impulse response of a material in the presence of an electric field as a mechanical system for which

$$\vec{x}(t) = \frac{-e/m}{(\omega^2 - \omega_0^2) + j\gamma\omega} E_0 e^{j\omega t}, \tag{21}$$

where $E_0$, $\omega_0$, $e$, $m$, and $\gamma$ are the electric field amplitude, the resonant frequency, the electric charge, the mass of the electron, and the damping constant, respectively [11]. We can multiply the quantity described in (21) be $e$ to get the dipole moment vector. We additionally multiply by the dipole density $N$ to get the bulk polarization. We see in (9) that the displacement field is equal to the added effect of the field and the polarization response of the material. However, we get a different value because we are not dealing in the plasmonic regime. The relative permittivity or the square of the index $n$ is expressed as

$$n^2(\omega) = \frac{\epsilon(\omega)}{\epsilon_0} = 1 + \frac{Ne^2/\epsilon_0 m}{(\omega^2 - \omega_0^2) + j\gamma\omega} \tag{22}$$

We can further imagine that there are a certain number of mechanical systems operating simultaneously. This means that for the $i$th system, there will be a certain fraction of atoms $f_i$ in that system; i.e., the effective dipole density for the $i$th system is $f_i N$.

$$n^2(\omega) = \sum_i^Z 1 + \frac{f_i\left(\frac{Ne^2}{m\epsilon_0}\right)}{(\omega^2 - \omega_i^2) + j\gamma_i\omega} = \sum_i^Z 1 + \frac{f_i\left(\frac{Ne^2}{m\epsilon_0}\right)}{\left(\frac{4\pi^2c^2}{\lambda^2} - \frac{4\pi^2c^2}{\lambda_i^2}\right) + \frac{j2\pi c\gamma}{\lambda}}$$
$$= \sum_i^Z 1 + \frac{f_i\left(\frac{Ne^2}{m\epsilon_0}\right)\lambda^2}{\left(4\pi^2c^2 - \frac{4\pi^2c^2\lambda^2}{\lambda_i^2}\right) + j2\pi c\gamma\lambda} = \sum_i^Z 1 + \frac{f_i\lambda_i^2\left(\frac{Ne^2}{m\epsilon_0}\right)\lambda^2}{(4\pi^2c^2\lambda_i^2 - 4\pi^2c^2\lambda^2) + j2\pi c\gamma\lambda_i^2\lambda} \tag{23}$$

where $c$ is the speed of light.

We then make the assumption that $\gamma$, the damping constant, goes to zero for the assumption of a simple harmonic oscillator.

$$n^2(\lambda) = \sum_i^Z 1 + \frac{f_i\lambda_i^2\left(\frac{Ne^2}{4\pi^2c^2m\epsilon_0}\right)\lambda^2}{(\lambda_i^2 - \lambda^2) + \frac{j\gamma\lambda_i^2}{2\pi c}\lambda} \underbrace{\approx}_{\gamma\to 0} \sum_i^Z 1 + \frac{f_i\lambda_i^2\left(\frac{Ne^2}{4\pi^2c^2m\epsilon_0}\right)\lambda^2}{(\lambda_i^2 - \lambda^2)} \tag{24}$$

We can make further simplifications to say that

$$n^2(\lambda) = \sum_i^Z 1 + \frac{G_i\lambda^2}{(\lambda_i^2 - \lambda^2)}, \tag{25}$$

where $\lambda_i$ is the $i$th resonant wavelength and the coefficient $G_i$ is the strength of that resonance. The coefficient $G_i$ is expressed as

$$G_i = \left(\frac{Ne^2}{4\pi^2c^2m\epsilon_0}\right)f_i\lambda_i^2 \tag{26}$$

These coefficients are known as Sellmeier coefficients [15]. These coefficients play a role in determining the group velocity of a particular wavelength in a fiber.

One of the misconceptions brought about by our knowledge of lasers is the "myth of monochromaticity." The light sources we use—even for optical fibers—are never fully monochromatic. Rather they emit a spectrum of frequencies (and wavelengths) concentrated around a particular center [16]. This spectrum will generate respectively scaled eigenmodes with their own group velocities and delays through the fiber. This causes pulse broadening even when only a single mode is propagating.

For a typical laser with a homogenous linewidth undergoing natural line broadening due to atomic decay, we can express the line-shape function as a typical Lorentzian for which

$$\phi_N(\nu) = \frac{1}{\pi} \frac{\Delta\nu_N/2}{(\nu - \nu_0)^2 + \left(\frac{\Delta\nu_N}{2}\right)^2}, \tag{27}$$

where $\Delta\nu_N$ is the FWHM frequency point or the natural linewidth [17]. This is the spectral profile of a typical laser centered around a frequency $\nu = \nu_0$. We can convert this line-shape function to one of the $k$-vector$k$ where

$$\phi_N(k) = \frac{1}{\pi} \frac{\Delta\nu_N/2}{\left(\frac{kc}{2\pi} - \frac{k_0 c}{2\pi}\right)^2 + \left(\frac{\Delta\nu_N}{2}\right)^2} = \frac{2\pi\Delta\nu_N/c^2}{(k - k_0)^2 + \left(\frac{\pi\Delta\nu_N}{c}\right)^2}. \tag{28}$$

where $k_0$ is the center spatial frequency.

The spectrum described in (28) shows that there is an uncountable infinity—i.e., a continuum—of $k$-vectors (and thus wavelengths) that is allowed.

In the case of a gas, the same can be done with the Boltzmann relation that is used to compute the statistical distribution of velocities (and thus $k$-vectors) under Doppler [18] broadening where the velocity component $v$ is normally distributed $\mathcal{N}(0, k_B T/m)$ while the energy $E$ undergoes a chi-squared distribution $\chi_1^2(E)$ as shown in (28):

$$P(v) = \sqrt{\frac{m}{2\pi k_B T}}\, e^{-\frac{mv^2}{2k_B T}} \underset{E=mv^2/2}{\leftrightarrow} P(E) = \frac{1}{\sqrt{4\pi k_B T}} \frac{e^{-\frac{E}{k_B T}}}{\sqrt{E}} \tag{29}$$

where $m, k_B, T$, and $v$ are the particle's mass, Boltzmann's constant, the temperature, and the velocity, respectively.

Because of the Doppler effect, the radiation will be shifted to a higher frequency $\nu$ which can be expressed as

$$\nu = \nu_0 \left(1 + \frac{v}{c}\right) \tag{30}$$

where $\nu_0$ is the center frequency.

With (30) in mind, we can rewrite (28) in terms of $\nu$ and convert to the $k$-domain like so

$$P(\nu) = \frac{c}{\nu_0} \sqrt{\frac{m}{2\pi k_B T}}\, e^{-\frac{mc^2(\nu - \nu_0)^2}{2\nu_0^2 k_B T}} \leftrightarrow P(k) = \frac{c}{k_0} \sqrt{\frac{m}{2\pi k_B T}}\, e^{-\frac{mc^2(k - k_0)^2}{2k_0^2 k_B T}} \tag{31}$$

What this means is that in the case of a realistic laser, dispersion within a confined medium will result even if the medium allows for only *one* mode to be activated. This is to say that Lorentzian homogenous broadening and Gaussian Doppler broadening *both* result in intramodal pulse broadening.

What is interesting is the counterintuitive connection between the frequency broadening and the temporal pulse broadening as their behaviors mirror each other. However, this is because we are dealing with *two* separate systems that happen to be connected. The first system is that of the laser linewidth and the decaying atoms and spontaneous thermal processes that diminish monochromaticity. These processes spread the spectrum out and a broader range of frequency (and wavelength) contribution to the spectrum. The second system is that of the delays accrued within the fiber-waveguide channel according to the wavelength contributions of the input spectrum. The input laser spectrum allows for a larger contribution to delays that will be "experienced" by the larger range of frequency and wavelength components, which *broaden* the temporal profile of the signal.

As noise is intrinsic to a lasing system so is dispersion when that system is directed into a medium; this is especially true in the context of confined or *bounded* media and their applications in modern communication networks.

## 8. Conclusion

Dispersion has been discussed in this chapter as it relates to electric fields in various media. We first begin with an understanding of electron responses to EM radiation; i.e., the light-matter interaction as understood in classical electromagnetics. When then used this concept to understand how the velocity of a signal through a medium is a function of the frequency of that signal.

Additionally, we explained and developed the concept of a group velocity with a simple analogy and then moved on to discuss that concept at length in various media. We contrasted group velocity with phase velocity and developed dispersion relations which contrasted the two.

We took the discussion of dispersive materials to light-confining structures. Optical waveguides and fibers function as dispersive materials not only because of their material properties but also because these materials are arranged. The arrangement of these materials—i.e., core-cladding indices, size, index distribution, etc.—all plays a role in the phenomenon of dispersion. This also influences the allowable forms of light that propagate in one of these confined media; the properties of the medium affect these modes of propagation. We additionally learned about ways in which this effect can be eliminated for engineering applications.

We finally ended this chapter with a discussion of effect of intramodally dispersive media on the frequency spectrum and time evolution of electric field signals. This was demonstrated by first dismantling the "myth of monochromaticity" and understanding light itself as a spectrum. This helped us better understand how a single mode can undergo dispersion with itself. It most importantly demonstrated the degree to which dispersion in inherent in EM propagation through all media.


## Author details

Emeka Ikpeazu
University of Virginia, Charlottesville, Virginia, United States of America

*Address all correspondence to: eikpeazu@gmail.com

## References

[1] Le Bourhis E. Appendix 1. Glass; Mechanics and Technology. 1st ed. Weinheim: Wiley-VCH; 2007

[2] Rao NN. Waveguide Principles. Fundamentals of Electromagnetics for Electrical and Computer Engineering. 1st ed. Upper Saddle River, NJ: Pearson/ Prentice Hall; 2009. pp. 290-338. DOI: 10.1002/978047974704.ch8

[3] Isaac N. Opticks. London: Royal Society; 1704. ISBN: 0-486-60205-2

[4] Staelin DH. Electromagnetics and Applications. Cambridge, MA: Department of Electrical Engineering and Computer Science, Massachusetts Institute of Technology; 2011

[5] Charles K. Introduction to Solid State Physics. 8th ed. Hoboken, NJ: Wiley; 2005

[6] Newman R, Tyler WW. Photoconductivity of germanium. In: Seitz F, Turnbull D, editors. Solid State Physics. Vol. 8. New York: Academic Press; 1959. pp. 49-107

[7] Rao NN. Waveguide Principles. Fundamentals of Electromagnetics for Electrical and Computer Engineering. 1st ed. Upper Saddle River, NJ: Pearson/ Prentice Hall; 2009. p. 142. DOI: 10.1002/978047974704.ch8

[8] Chen LF, Ong CK, Neo CP, Varadan VV, Varadan VK. Microwave Electronics: Measurement and Materials Characterization. 1st ed. Chichester: Wiley; 2004

[9] Rao NN. Waveguide Principles. Fundamentals of Electromagnetics for Electrical and Computer Engineering. 1st ed. Upper Saddle River, NJ: Pearson/ Prentice Hall; 2009. p. 143. DOI: 10.1002/978047974704.ch8

[10] Engineering ToolBox. Relative Permittivity—The Dielectric Constant [online]. 2010. Available from: https:// www.engineeringtoolbox.com/relative-permittivity-d_1660.html [Accessed: 09 October 2019]

[11] Pollock CR. Dispersion in Waveguides. Fundamentals of Optoelectronics. 1st ed. Chicago: Irwin; 2003

[12] Mitschke F. Fiber Optics: Physics and Technology. Heidelberg, Dordrecht, London, New York: Springer; 2009. p. 22

[13] Mitschke F. Fiber Optics: Physics and Technology. Heidelberg, Dordrecht, London, New York: Springer. 2009. p. 21

[14] Pollock CR. Graded-index Waveguides. Fundamentals of Optoelectronics. 1st ed. Chicago: Irwin; 2003

[15] Sellmeier W. Über die durch die Aetherschwingungen erregten Mitschwingungen der Körpertheilchen und deren Rückwirkung auf die ersteren, besonders zur Erklärung der Dispersion und ihrer Anomalien (II. Theil). Annalen der Physik und Chemie. 1872;**223**(11):386-403

[16] Kasap SO. Optoelectronics and Photonics–Principles and Practices. 1st ed. Upper Saddle River, NJ: Prentice Hall; 2001. p. 78

[17] Peach G. Theory of the pressure broadening and shift of spectral lines. Advances in Physics. 1981;**30**(3):367-474

[18] Siegman AE. Lasers. 1st ed. Mill Valley, CA: University Science Books; 1986

Chapter 3

# The Electrical Properties of Soils with Their Applications to Agriculture, Geophysics, and Engineering

*Walter Gustavo Fano*

## Abstract

The electric properties of the soils are very important for several sciences like telecommunications, electrical engineering, geophysics, and agriculture. There are semiempirical dielectric models for soils, which represent the real and imaginary part of the dielectric permittivity as the function of the frequency. The measurement methods to obtain the dielectric properties of soils are described for different bands of frequencies from some kHz to several GHz. The parallel plate capacitors are widely used to measure dielectric properties. The transmission line method of a coaxial transmission line can be used in frequency domain and time domain. The time domain technique with transmission lines is usually called time-domain reflectometry (TDR), because it is based on the voltage measurement as a function of time of pulses. The frequency domain technique with transmission lines is based on the reflection coefficient measurement of the transmission line. The transmission line method is described with short load and open-circuit load because it is useful in obtaining the characteristic impedance and the electric permittivity of the media inside.

**Keywords:** soil, dielectric properties, permittivity, models, measurement method

## 1. Introduction

The knowledge of electrical properties of soils in physics and electrical engineering are important for many applications. The long-distance electromagnetic telegraph systems from 1820 are used, with two or more wires to carry the signal and the return currents. It was discovered that the earth could be used as a return path to complete the circuit, making the return wire unnecessary [1]. However, during dry weather, the earth connection often developed a high resistance, requiring water on the earth electrode to enable the telegraph to ring [1].

An important radio propagation and engineering problem has been solved in 1909 by A. Sommerfeld. He has solved the general problem of the effect of the finite conductivity of the ground on the radiation from a short vertical antenna at the surface of a plane earth. The surface wave propagation is produced over real ground for the medium frequency AM radio service, where the attenuation of the electric field depends on the dielectric properties of the soil, mainly of the dielectric losses [2]. Considering the word "Soil" means the uppermost layer of the earth's crust, it contains the organic as well as mineral matter. From 1936 up to 1941, Norton, Van der Pol, and Bremmer made the computation of the field strengths at distant points on the flat and spherical Earth's surface [3, 4].

In agriculture applications, the electrical resistivity methods have been introduced by Conrad Schlumberger in France and Frank Wenner in the United States, for the evaluation of ground electrical resistivity. In saline soils, the electric conductivity measured is high, and the effects of salinity are manifested in the loss of stand, reduced rates of plant growth, reduced yields, and in severe cases, total crop failure [5].

The applications like the protection of electrical generating plant are necessary to provide earth connections with low electrical resistance. The radio transmitting and receiving stations for broadcasting is generally covered by radiation transmitted directly along the ground [6]. In electrical engineering, "ground" is the reference point in an electrical circuit from which voltages are measured.

For archeology, geophysics, engineering, and military applications, the so-called ground-penetrating radar (GPR) is a technique widely used. The radar signal is an electromagnetic wave that propagates through the earth, and its signal is reflected when an object appears or there is a change in the properties of the earth. In order to determine the depth of an object under the ground, it is necessary to know the electrical properties of the soil [7].

## 2. Fundamental concepts

The equations that relate the electric field (E) and magnetic field (H) are based on the electromagnetic theory formulated by James Clerk Maxwell in 1864, whose validity has allowed great advances in diverse areas, such as telecommunications, electricity, electronics, and materials [8].

Regarding the behavior of the materials under the action of an electric field, in the conductive materials, the charges can move freely, meaning that the electrons are not associated with an atomic nucleus. In the case of dielectric materials, the charges are associated with an atom or specific molecule [9]. There are two main mechanisms where the electric field distorts the distribution of charge in a dielectric, stretching and rotation. The relationship between the electric dipole moment inducted under the action of an applied electric field is called atomic electric polarizability $\vec{p}$ and can be written as

$$\vec{p} = \alpha\vec{E} \tag{1}$$

In a material with an applied electric field, a convenient definition is to consider the contributions of the dipole moment per unit volume; this parameter is called polarization, which is a macroscopic definition instead of a molecular or atomic definition [9, 10]:

$$\vec{P} = \frac{\Delta\vec{p}}{\Delta v} \tag{2}$$

It is evident that the contributions of the electric dipole moment in a volume element $\Delta v$ are given by the sum of the microscopic contributions $\overrightarrow{p_m}$; therefore you can write:

$$\Delta\vec{p} = \sum_m \overrightarrow{p_m} \tag{3}$$

And the polarization

$$\vec{P} = \frac{\sum_m \overrightarrow{p_m}}{\Delta v} \tag{4}$$

In **Figure 1**, the external electric field applied to a dielectric material and the resulting polarization can be observed.

From the macroscopic point of view in most of the dielectric material, when the electric field is canceled, the polarization in the material will be nullified. In addition, the polarization of the material will vary as the electric field varies, i.e., $\overrightarrow{P(E)} = \chi(E)\vec{E}$. The variable $\chi(E)$ is called electrical susceptibility of the material, and when the electric field applied to the dielectric is not high powered, the relationship between the polarization and the electric field will be linear, as is the case of a soil [10]:

$$\vec{P} = \chi\vec{E} \tag{5}$$

It is convenient to define the electric displacement, because it allows to relate by means of the Gaussian law with the free charges; therefore

$$\vec{D} = \epsilon_0 E + \vec{P} \tag{6}$$

Then

$$\vec{D} = \epsilon_0 E + \chi\vec{E} \tag{7}$$

The electrical permittivity is defined as the relationship between the electric displacement vector $\vec{D}$ and the electric field vector $\vec{E}$, and thus

$$\vec{D} = \epsilon\vec{E} \tag{8}$$

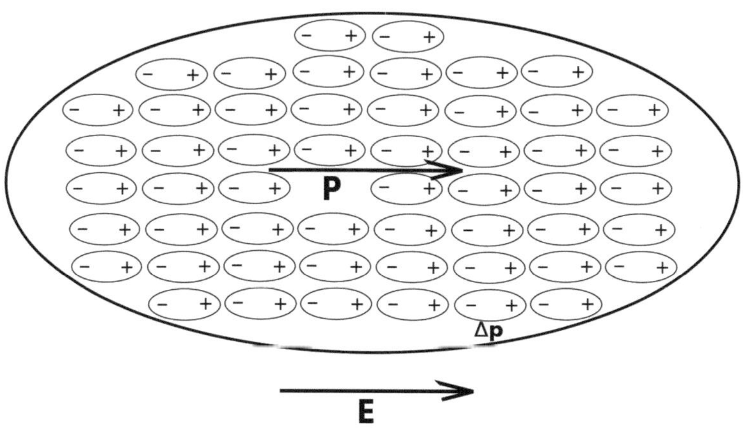


**Figure 1.**
*Polarization applying external electric field E, to a dielectric material.*

Result

$$\epsilon = \epsilon_0 + \chi \tag{9}$$

It is convenient to define [11]:

$$\chi = \epsilon_0 \chi_r \tag{10}$$

Result

$$\epsilon = \epsilon_0 (1 + \chi_r) \tag{11}$$

The electric properties of the material are completely defined by means of $\varepsilon$ or $\chi$ [10].

In problems with electromagnetic fields, four vectors are defined: E and B; D and H. These vectors are assumed to be finite throughout the entire field, and at all ordinary points to be continuous functions of position and time, with continuous derivatives [12]. The constitutive relations link the vectors of the fields $\vec{B}$ with $\vec{H}$ and $\vec{D}$ with $\vec{E}$, usually dependent on the frequency [8]:

$$\begin{aligned} \vec{B} &= \mu(\omega)\vec{H} \\ \vec{D} &= \epsilon(\omega)\vec{E} \\ \vec{J} &= \sigma(\omega)\vec{E} \end{aligned} \tag{12}$$

For the electromagnetic propagation in soils, the parameters $\mu$, $\epsilon$, and $\sigma$ must be determined. The soils are usually nonmagnetic media; therefore the magnetic permeability will be that of the vacuum $\mu = \mu_0$, and the variables to be determined will be the electric permittivity $\epsilon$ and the electric conductivity $\sigma$.

The electrical resistivity obtained by soil mapping exhibits a large range of values from $1\Omega/m$ for saline soil to several $10^5\Omega/m$ for dry soil overlaying crystalline rocks [13]. In **Figure 2**, the resistivity of different soils can be observed. The electrical conductivity is related to the particle size by the electrical charge density

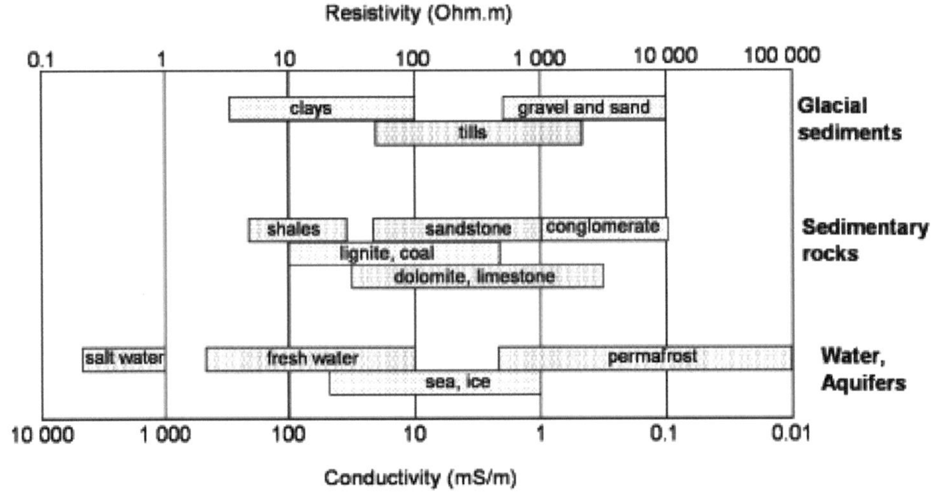


**Figure 2.**
*Table of electric resistivity $[\Omega/m]$ and electric conductivity $[\sigma/m]$ of soils (Ref. Samoulian et al.) [13].*

at the surface of the solid constituents, like in clay soil; the electrical charges located at the surface of the clay particles lead to greater electrical conductivity than in coarse-textured soils [13].

There are evidences that for compacted soils of clay, it exhibits an anisotropic behavior in the resistivity measured in the horizontal and vertical directions [14].

The literature contains the measurement of the dielectric properties of soils at different frequencies with slotted lines and time-domain reflectometry (TDR) methods [15].

The measured variations of the electric permittivity of soils with fractions of sand, silt, and clay and with volumetric moisture content have been studied for frequency of 440 MHz used by the radar observations [16].

The coaxial probe technique terminated in the material under test has been used to measure the dielectric properties of the vegetation. The dielectric data reported are based on measurements of the amplitude and phase of the reflection coefficient of a coaxial probe [17, 18].

## 3. Transmission line fundamentals

The transmission line method has been used to measure the dielectric properties [19, 20]. These transmission lines are coaxial, quasi-coaxial, and two-wire transmission lines. Consider a transmission line with a homogenous dielectric material inside, and the propagation is transverse electromagnetic mode (TEM), where the electric and magnetic field are perpendicular to the propagation direction; this can be observed in **Figures 3** and **4**.

The separation between the conductive cylinders that form the coax transmission line should be much lower than the wavelength of the signal that propagates, so the transmission line will not be affected by the propagation modes of high orders, such as the $TE_{11}$ [19].

Coaxial transmission lines are widely used for the transmission of radio-frequency signals and its application in radiocommunications and for broadcasting [21]. The transmission lines allow the connection between a generator or emitter and a load or antenna. The air coaxial transmission line consists of two cylindrical conductors, with air between both conductors. These metallic conductors are those that impose the boarder conditions that must comply with the electric and magnetic fields of the electromagnetic wave that travel inside the line. The coaxial

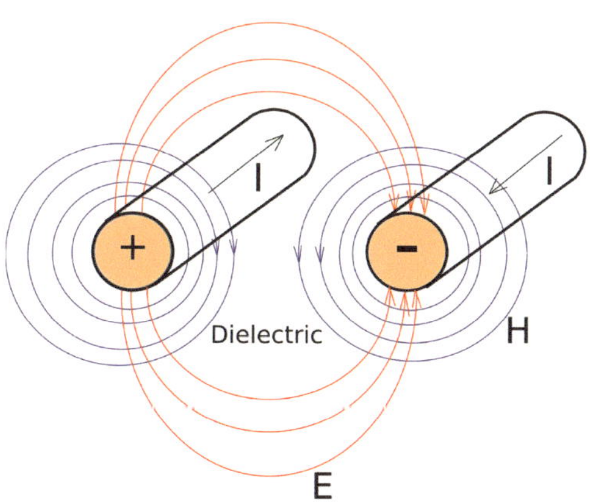


**Figure 3.**
*Section of the two-wire transmission line with the electric and magnetic fields.*

transmission lines are used to measure the electrical properties of a dielectric material located inside the coaxial transmission line, as shown in **Figure 4**.

By analyzing the circuit model of a transmission line, the currents and voltages that propagate along it can be determined, using the circuit theory [22]. The equivalent circuit model of a transmission line can be seen in **Figure 5**. According to the equivalent circuit model of a transmission line, the characteristic impedance $Z_0$ and the propagation constant $\gamma$ can be expressed thus [21]:

$$Z_0 = \sqrt{\frac{R + j\omega L}{G + j\omega C}} \tag{13}$$

$$\gamma = \sqrt{(R + j\omega L)(G + j\omega C)} \tag{14}$$

where R is the series resistance per unit length $[\Omega/m]$; L is the series inductance per unit length $[H/m]$; C is the parallel capacity per unit length $[F/m]$; G is the parallel conductance per unit length $[S/m]$.

If the transmission line has no losses, it means that R = 0 and G = 0; then the characteristic impedance can be reduced as follows:

$$Z_0 = \sqrt{\frac{L}{C}} \tag{15}$$

$$\gamma = j\omega\sqrt{LC} \tag{16}$$

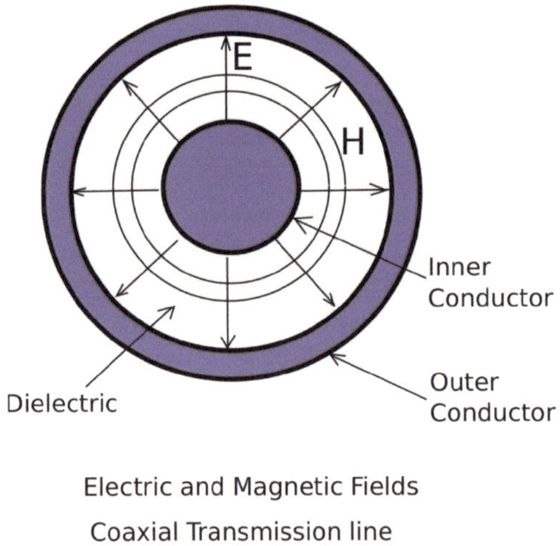


**Figure 4.**
*Section of coaxial transmission lines and the electric and magnetic fields.*

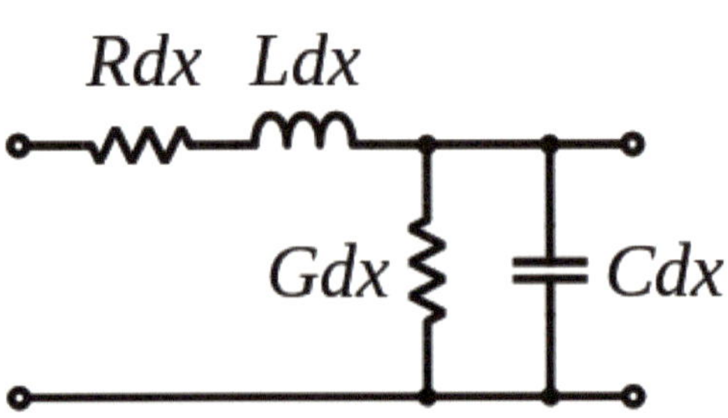


**Figure 5.**
*Distributed parameters of the transmission line.*

The input impedance of a transmission line, with a material inside considering the material with dielectric losses, can be expressed thus [23]:

$$Z_i = Z_0 \frac{Z_L + Z_0 \; tanh \; (\gamma l)}{Z_0 + Z_L \; tanh \; (\gamma l)} \tag{17}$$

where $\gamma$ is the propagation constant $[1/m]$; $Z_L$ is the load impedance $[\Omega]$; $l$ is the length of the transmission line from the load $[m]$; $Z_0$ is the characteristic impedance of the transmission line $[\Omega]$.

## 4. Time-domain measurement method of dielectric permittivity and conductivity of soils (TDR)

The time-domain reflectometry (TDR) is a well-known technique used to find the interruption point of the transmission lines in a CATV installation and is also useful to determine the dielectric permittivity (see **Figure 6**).

The time-domain reflectometry uses a step generator and an oscilloscope; a fast edge is launched into the transmission line under investigation, where the incident and reflected voltage waves on the transmission line are monitored by the oscilloscope. This method shows the losses and the characteristic impedance of the line: resistive, inductive, or capacitive [25]. The TDR method is based on the velocity of the electromagnetic wave that propagates through the soil, and the velocity of the wave depends on the water content of the soil. If a pulse is applied to a no-loss transmission line, the time domain graphic can be shown like in **Figure 7**. Considering the soil like a nonmagnetic media with low dielectric loss is [26, 27]:

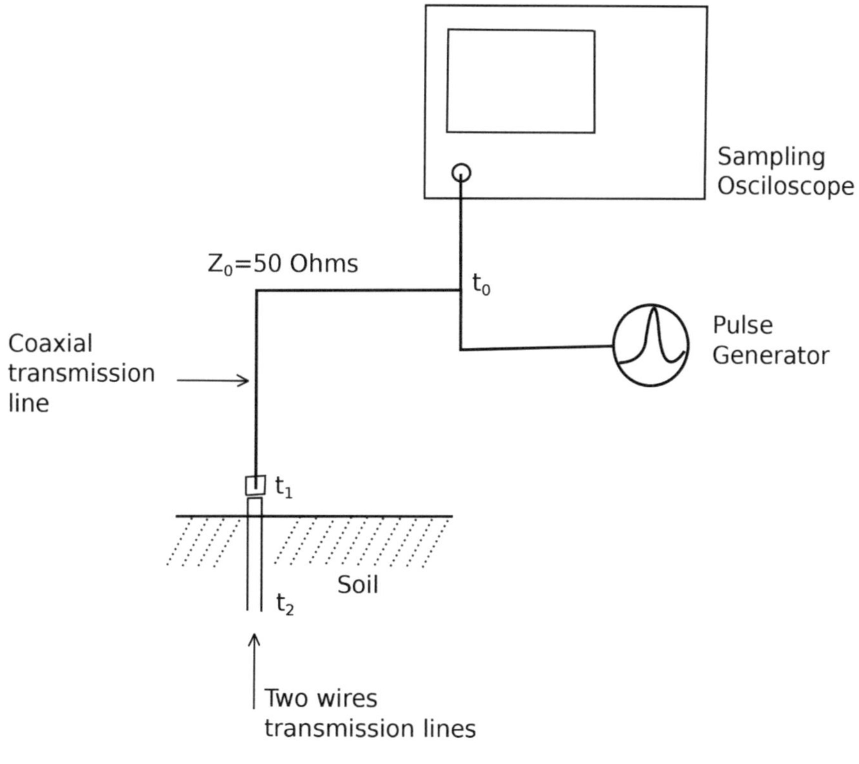


**Figure 6.**
*Setup of the dielectric measurement by the TDR method [24].*

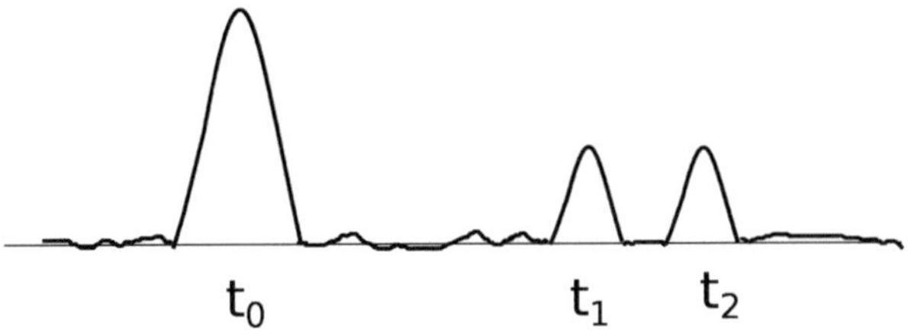


**Figure 7.**
*Propagation of the pulses in the time domain graphic with dielectric air [24].*

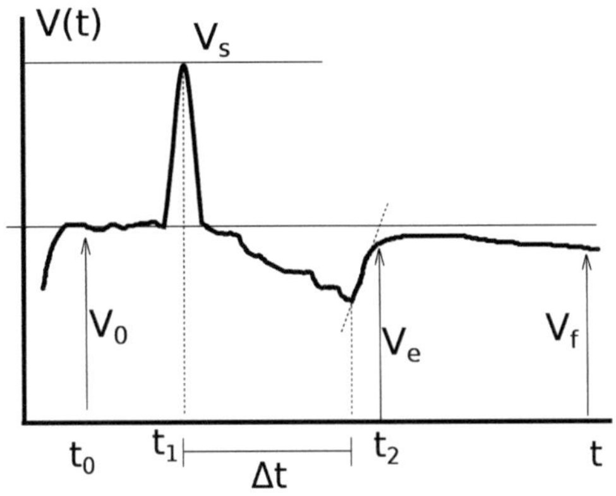


**Figure 8.**
*Picture of the voltage as a function of time for the probe is in the soil [26].*

$$v = \frac{c}{\sqrt{\epsilon}} \tag{18}$$

where $c[m/s]$ is the light velocity; $\epsilon$ is the electric permittivity of the soil under test.

The time interval $\Delta t$ between the received pulse and incident pulse can be observed in **Figure 8**, and the velocity can be expressed thus:

$$v = \frac{c}{\sqrt{\epsilon}} = \frac{2L}{\Delta t} \tag{19}$$

where $L$ is the probe length.

Then

$$\epsilon = \left(\frac{c\Delta t}{2L}\right)^2 \tag{20}$$

Usually the transmission line probes have a minimum length of 15 cm, because the incident electromagnetic wave takes a time of 1 ns in air in order to return to the input of the line. This time is too short to be measured.

The conductivity of the soil can be determined computing the reflected pulses in the probe in the time domain graphic (see **Figure 8**) [26, 28]. Numerous methods have been proposed by researchers; one of these is the procedure of Dalton et al. (1984) [26]:

$$\sigma_{dalton} = \left(\frac{\sqrt{\epsilon}}{120\pi L}\, ln\left(\frac{V_f}{V_e - V_f}\right)\right) \tag{21}$$

where $V_0$ is the amplitude of the TDR pulse; $V_s$ is the amplitude after reflection from the start of the probe; $V_e$ is the amplitude after reflection from the end of the probe; $V_f$ is the reflected signal after a very long time.

Also the conductivity can expressed thus [29]:

$$\sigma = \frac{K}{Z_0}\left(\frac{1-\rho}{1+\rho}\right) \tag{22}$$

where:

$$\rho = \frac{V_f - V_0}{V_0}$$

$Z_0$ is the characteristic impedance of the transmission line.

$K$ is a geometric constant of the probe, and it is experimentally determined by immersing the probe in solutions of known electrical conductivity Nt at temperature:

$$K = EC \cdot RL \cdot ft^{-1}\,(m)^{-1}$$

Temperature correction $fT = 1 + KT(T - 25)$ $KT$ depends on the used solution. $KT = 0.0191$ for a 0.01 M KCl solution.

## 5. Measurement method of dielectric permittivity and conductivity of soils in frequency domain

This method is based on the measurement of the reflection coefficient by means of the vector network analyzer (VNA) on the frequency domain of a coaxial transmission line in the soil; this can be observed in **Figure 9** [30–32].

The vector network analyzer can measure the scattering coefficient of a two-port passive network where the reflection coefficient in voltage $\Gamma$ is $S_{11}$ [34]. The probe impedance with the material inside is related with the reflected coefficient $\Gamma$ [31]:

$$\frac{Z_p}{Z_0} = \frac{1+\Gamma}{1-\Gamma} \tag{23}$$

where $Z_p$ is the impedance of the probe at the load and $Z_0$ is the characteristic impedance of the probe.

The impedance of the probe can be calculated thus:

$$Z_p = \frac{2\,log\,\frac{B}{A}}{\sqrt{c\epsilon *}} cotangh \frac{\omega L\sqrt{\epsilon *}}{c} j \tag{24}$$

where $L$ is the electric length of the probe; $c$ is the speed of light; $L$ is the coaxial probe of length; $A$ is the inner diameter; $B$ is the outer diameter.

Then the complex electric permittivity for frequencies lower than 50 MHz can be approximated thus [31]:

$$\epsilon = \left(\frac{\frac{L}{2\,log\,\frac{B}{A}}}{\frac{1}{Z_T \omega j} - C_s} + \frac{\omega^2 L^2}{3c^2}\right)^{-1} \tag{25}$$

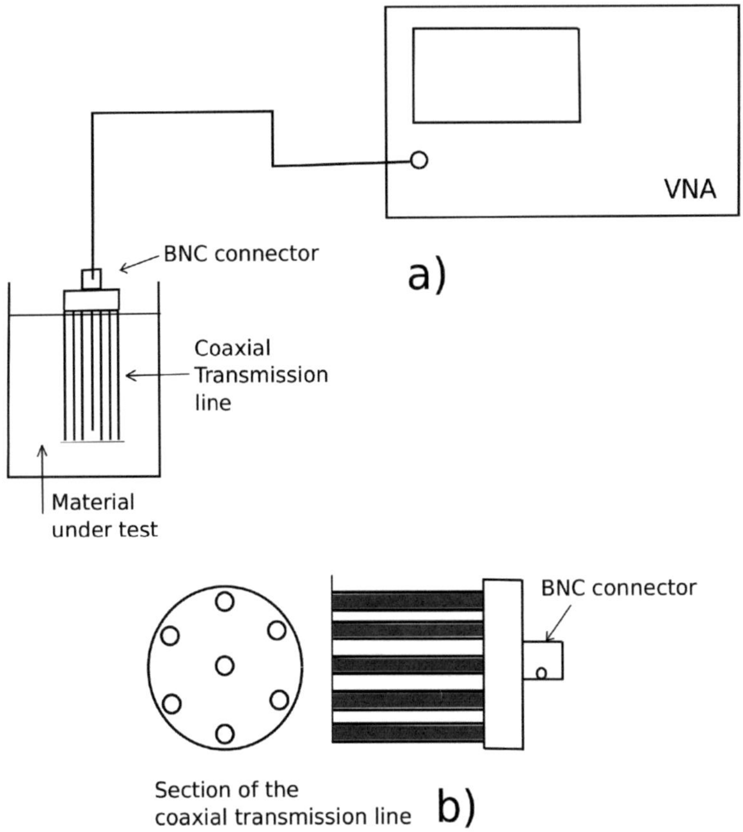


**Figure 9.**
*Dielectric measurement by the coaxial transmission line method. (a) Setup of the measurement experiment; (b) section of the transmission line and the material under test [33].*

## 6. Dielectric measurement by the characteristic impedance of a transmission line in frequency domain

Some references of these measurement methods by means of characteristic impedance have been developed [35, 36]. This methods is shown in **Figure 10**.

The input impedance can be computed by Eq. (17) for two different loads' impedance:

(a) Open circuit in the load $Z_L \to \infty$

$$Z_i|_{Z_L=\infty} = \frac{Z_0}{tanh\ (\gamma l)} = Z_0\ tanh\ ^{-1}(\gamma l) \tag{26}$$

(b) Short circuit in the load $Z_L = 0$

$$Z_i|_{Z_L=0} = Z_0\ tanh\ (\gamma l) \tag{27}$$

where, in general, the material inside the transmission line could be a dielectric loss; the propagation constant can be written thus:

$$\gamma = \alpha + j\beta \tag{28}$$

where $\alpha$ is the attenuation constant [Neper/m] and $\beta$ is the phase constant [rad/m].

Using Eqs. (26) and (27)

$$\gamma|_{Z_L \to \infty} = \frac{1}{l} atanh\left(\frac{Z_0}{Z_i}\right) \tag{29}$$

$$\gamma|_{Z_L = 0} = \frac{1}{l} atanh\left(\frac{Z_i}{Z_0}\right) \tag{30}$$

Using the relation between $ln\ (x)$ and $th^{-1}(x)$

$$\gamma|_{Z_L \to \infty} = \frac{1}{2l}\ ln\ \frac{Z_0 + Z_i}{Z_0 - Z_i} \tag{31}$$

$$\gamma|_{Z_L = 0} = \frac{1}{2l}\ ln\ \frac{Z_i + Z_0}{Z_i - Z_0} \tag{32}$$

The argument of the ln

$$\gamma|_{Z_L \to \infty} = \frac{1}{2l}\ ln\ \left(\rho_1 e^{\phi_1}\right) \tag{33}$$

$$\gamma|_{Z_L = 0} = \frac{1}{2l}\ ln\ \left(\rho_2 e^{\phi_2}\right) \tag{34}$$

Replacing the Ln

$$\gamma|_{Z_L \to \infty} = \frac{1}{2l}\ ln\ \left(\rho_1 e^{\phi_1}\right) = \frac{1}{2l}\left[\ ln\ (\rho_1) + \ ln\ \left(e^{j\phi_1}\right) + j2k\pi\right]\ k = 0, 1, 2, 3.... \tag{35}$$

$$\gamma|_{Z_L = 0} = \frac{1}{2l}\ ln\ \left(\rho_2 e^{\phi_2}\right) = \frac{1}{2l}\left[\ ln\ (\rho_2) + \ ln\ \left(e^{j\phi_2}\right) + j2k\pi\right]\ k = 0, 1, 2, 3.... \tag{36}$$

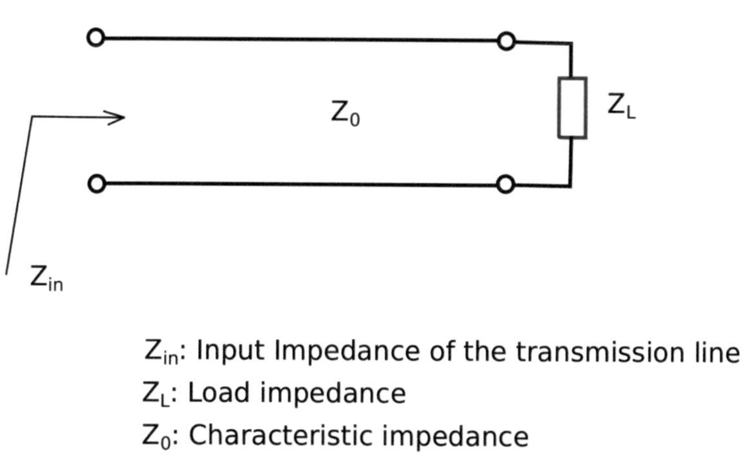


**Figure 10.**
*Input impedance of the transmission line for $Z_L = 0$ and $Z_L \to \infty$.*

Then the propagation constant can be written thus:

$$\begin{aligned} \alpha|_{Z_L\to\infty} &= \frac{1}{2l}\ ln\ (\rho_1) \\ \beta|_{Z_L\to\infty} &= \frac{1}{2l}[\phi_1 + 2k\pi]\ k = 0, 1, 2, 3.... \\ \alpha|_{Z_L=0} &= \frac{1}{2l}\ ln\ (\rho_2) \\ \beta|_{Z_L=0} &= \frac{1}{2l}[\phi_2 + 2k\pi]\ k = 0, 1, 2, 3.... \end{aligned} \tag{37}$$

By these last equations, the attenuation constant and the phase constant can be calculated with $Z_i|_{Z_L=0}$ or $Z_L \to \infty$.

## 6.1 Determining ϵ and σ from the propagation constant

The propagation constant $\gamma$ can be written thus [11]:

$$\gamma = \sqrt{j\omega\mu(\sigma + j\omega\epsilon)} \tag{38}$$

From Eqs. (38) and (28)

$$\gamma^2 = j\omega\mu(\sigma + j\omega\epsilon') = -\omega^2\mu\epsilon' + j\omega\mu\sigma \tag{39}$$

$$\gamma^2 = \alpha^2 - \beta^2 + 2j\alpha\beta = (\alpha^2 + \beta^2)e^{j2atan\left(\frac{\beta}{\alpha}\right)} \tag{40}$$

Equating real and imaginary parts of $\gamma^2$ of Eqs. (39) and (40)

$$\begin{aligned} -\omega^2\mu\epsilon' &= (\alpha^2 + \beta^2)\ cos\ \left(2atan\left(\frac{\beta}{\alpha}\right)\right) \\ \omega\mu\sigma &= (\alpha^2 + \beta^2)\ sin\ \left(2atan\left(\frac{\beta}{\alpha}\right)\right) \end{aligned} \tag{41}$$

$\epsilon$ and $\sigma$ can be obtained:

$$\begin{aligned} \epsilon' &= \frac{(\alpha^2 + \beta^2)\ cos\ \left(2atan\left(\frac{\beta}{\alpha}\right)\right)}{-\omega^2\mu} \\ \sigma &= \frac{(\alpha^2 + \beta^2)\ sin\ \left(2atan\left(\frac{\beta}{\alpha}\right)\right)}{\omega\mu} \end{aligned} \tag{42}$$

Another expression of $\epsilon$ and $\sigma$ is using Eqs.(39) and (40):

$$\begin{aligned} \gamma^2 &= j\omega\mu(\sigma + j\omega\epsilon') \\ \gamma^2 &= \alpha^2 - \beta^2 + 2j\alpha\beta \end{aligned} \tag{43}$$

Equating real and imaginary part of Eq. (43)

$$\begin{aligned} \omega\mu\sigma &= 2\alpha\beta \\ -\omega^2\mu\epsilon &= \alpha^2 - \beta^2 \end{aligned} \tag{44}$$

Results

$$\sigma = \frac{2\alpha\beta}{\omega\mu}$$
$$\epsilon = \frac{\beta^2 - \alpha^2}{\omega^2\mu} \quad (45)$$

The series resistance of the conductor of the coaxial transmission line used is $R_{in} \cong 10^{-3}\Omega/m$ $(f = 100MHz)$ $Z_L = 0$. This resistance can be neglected, because it has no effect in the measured attenuation constant $\alpha$.

### 6.2 Measurement procedure of the $\epsilon$ and $\sigma$

1. The input impedances are measured for a load impedance at short circuit $Z_L = 0$ and open circuit $Z_L \to \infty$ (see Eqs. (26) and (27)).

2. The attenuation constant $\alpha$ and phase $\beta$ are calculated with the equations found ad hoc (see Eq. (37)).

3. The electric permittivity $\epsilon$ and the electrical conductivity $\sigma$ are calculated using the equations found in Section 6.1.

In this way, a practical method of measurement is available to determine the parameters of dielectric materials, using coaxial transmission lines, in the frequency range from 1 to 1000 MHz. A problem that appears when measuring dielectric materials is the connector that establishes the link between the coaxial transmission line and the vector impedance meter. A systematic error in the impedance measured is introduced.

Therefore, the study and correction of the mentioned error in the section will be carried out.

### 6.3 Transmission lines used

Three coaxial transmission lines of General Radio (GR) Type 874, with air dielectric, have been used with a length of 100, 200, and 300 mm. The main characteristics of the General Radio coaxial transmission lines, type 874, are the following:

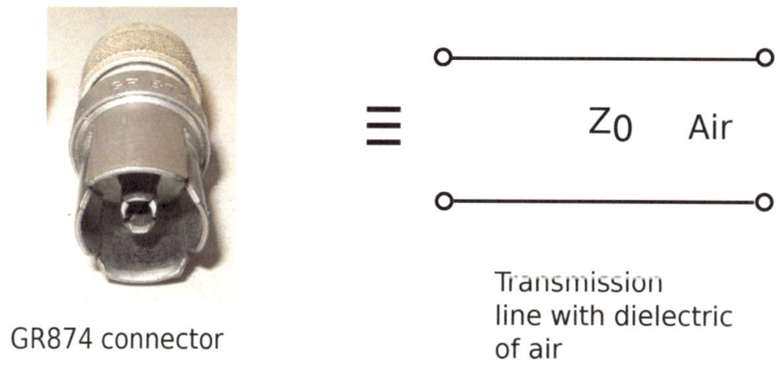


**Figure 11.**
*N connector and its equivalent of a transmission line with dielectric of air.*

Characteristic impedance $Z_0 = 50\Omega$.
Input and output connector GR874

$$r_1 = 12 \cdot 10^{-3}[m]$$
$$r_2 = 5.2 \cdot 10^{-3}[m]$$
$$\sigma \cong 5.810^7[S/m]$$

## 6.4 Correction error produced by the connector of the transmission line

It is important to perform the correction of the impedance introduced by the connector of the transmission line used. This connector is shown in **Figure 11**, and it is composed by a dielectric of very low dielectric losses and has a length of 10 mm (**Figure 12**). The characteristic impedance of the connector is practically $Z_0 = 50\Omega$ with no losses [36]:

$$\gamma = \alpha + j\beta \cong j\beta \tag{46}$$

The input impedance to the connector can be written thus:

$$Z_i = Z_{0con} \frac{Z_L + jZ_{0con} tg(\beta x)}{Z_{0con} + jZ_L tg(\beta x)} \tag{47}$$

where $x$ is the length of the transmission line [m].

The electric permittivity of the dielectric of the connector is unknown; then it is easy to assume a transmission line with air equivalent to the connector with $\epsilon = \epsilon 0$, $\mu = \mu_0$, and $Z_{0con} = 50\Omega$.

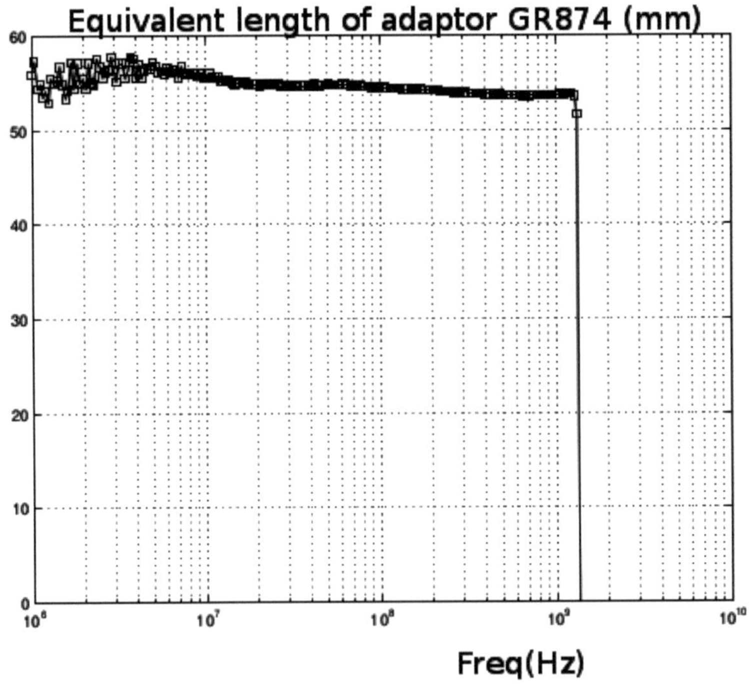


**Figure 12.**
*Equivalent length of the transmission line of the connector GR874.*

Considering the connector with no losses

$$\alpha = 0$$
$$\beta = \omega_0 \sqrt{\mu_0 \epsilon_0} \tag{48}$$

Then the input impedance of the connector with $Z_L = 0$ is

$$Z_i|_{Z_L=0} = jZ_{0con} tg(\beta x) \tag{49}$$

where $x$ is the length of the connector of the equivalent transmission line.
The length $x$ of this equivalent transmission line can be written thus:

$$x = \frac{1}{\beta} tg^{-1} \left( \frac{Z_{icon}|_{Z_L=0}}{jZ_{0con}} \right) \tag{50}$$

## 6.5 Results and discussion

### *6.5.1 Method of measurement*

The experimental results of the electric conductivity and the dielectric permittivity measurement of the dry sand can be observed in **Figures 13** and **14**. In **Figure 13**, the electric conductivity as a function of the frequency, by means of the capacitive method, and the three types of transmission line lengths have been measured: $L$ = 100, 200, and 300 mm; the convergence of all measurements are evident.

In **Figure 14**, the relative electric permittivity as a function of the frequency, by means of the capacitive method, and the three types of transmission line lengths

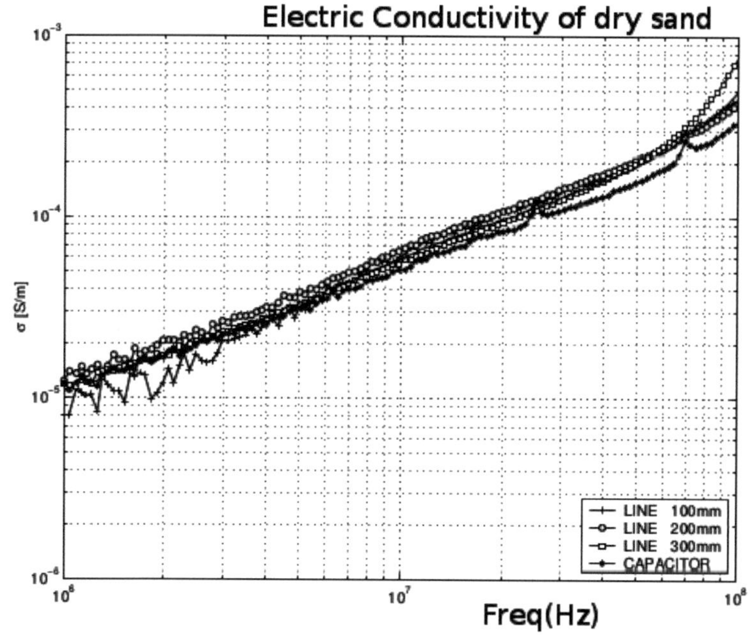


**Figure 13.**
*Electric conductivity as a function of the frequency for dry sand samples, using a capacitive method and three transmission lines: 100, 200, and 300 mm.*

have been measured: $L = 100$, 200, and 300 mm; there is a convergence of all measurements. It is important to note that the shorter transmission line has a wider bandwidth of measurement. The transmission line length of $L = 300mm$ shows the useful results up to 30 MHz; with a TL length of $L = 200mm$, the useful results are up to 50 MHz; and with a TL length of $L = 100mm$, the useful results are up to 300 MHz.

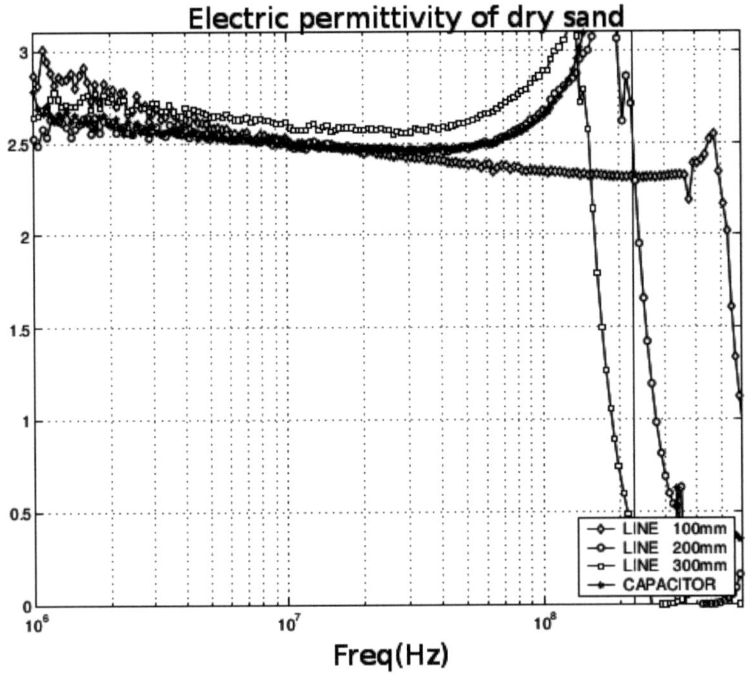


**Figure 14.**
*The relative dielectric permittivity as a function of the frequency for dry sand samples has been measured, using a capacitive method and three transmission lines: 100, 200, and 300 mm.*

| freq. (MHz) | Capacitor | T. line 100 mm | T. line 200 mm | T. line 300 mm | Expected value | Std. dev |
|---|---|---|---|---|---|---|
| 1 | 2.80 | 2.86 | 2.65 | 2.65 | 2.74 | 0.034 |
| 2 | 2.60 | 2.76 | 2.62 | 2.75 | 2.68 | 0.021 |
| 3 | 2.60 | 2.70 | 2.60 | 2.70 | 2.65 | 0.010 |
| 5 | 2.50 | 2.61 | 2.57 | 2.65 | 2.58 | 0.012 |
| 7 | 2.50 | 2.54 | 2.55 | 2.62 | 2.55 | 0.0075 |
| 10 | 2.50 | 2.51 | 2.50 | 2.60 | 2.52 | 0.0073 |
| 20 | 2.40 | 2.45 | 2.47 | 2.57 | 2.47 | 0.0153 |
| 30 | 2.40 | 2.42 | 2.46 | 2.57 | 2.46 | 0.0173 |
| 50 | 2.40 | 2.40 | 2.46 | — | 2.41 | 0.0028 |
| 70 | — | 2.37 | — | — | — | — |
| 100 | — | 2.35 | — | — | — | — |

**Table 1.**
*Relative electric permittivity of dry sand.*

| freq. (MHz) | Capacitor | T. line 100 mm | T. line 200 mm | T. line 300 mm | Expected value | Std. dev |
|---|---|---|---|---|---|---|
| 1 | 1 | 1.1 | 1.3 | 1.2 | 1.15 | 0.13 |
| 2 | 1.6 | 2.0 | 1.7 | 1.7 | 1.75 | 0.17 |
| 3 | 2.1 | 2.6 | 2.3 | 2.2 | 2.30 | 0.22 |
| 5 | 3.2 | 3.1 | 3.3 | 3.1 | 3.20 | 0.096 |
| 7 | 4.4 | 5 | 4.3 | 4 | 4.40 | 0.42 |
| 10 | 6.0 | 6.5 | 5.5 | 5.1 | 5.80 | 0.6 |
| 20 | 10 | 11 | 9.4 | 8.4 | 9.70 | 1.08 |
| 30 | 14.4 | 14.6 | 12.7 | 11.1 | 13.20 | 1.60 |
| 50 | 21.9 | 21.5 | 20.8 | 16 | 20.00 | 2.7 |

**Table 2.**
*Electric conductivity of dry sand* $\sigma\left[\frac{S}{m}10^{-5}\right]$.

The expected value of the dielectric permittivity measured for the dry sand by means of a parallel plate capacitor and the three transmission lines used are shown in **Table 1**. The standard deviation of the three measurements shows a good agreement up to the vicinity of the resonant frequency of each transmission line. In **Table 2**, the electric conductivity of the dry sand can be observed. These curves have the same slope and show a good convergence.

### *6.5.2 Applications*

The values of the electrical conductivity and the electrical permittivity are very useful to evaluate the propagation of surface waves in real ground, where the attenuation depends mostly on the conductivity of the soil. Such is the case that AM transmitters include radials, which consist of metallic conductors, placed at the base of the monopole antenna to increase conductivity, and in this way the losses due to Joule effect on the earth's surface are reduced. When the conductivity of the soil is perfect, the electric field vector that propagates will be perpendicular to the earth's surface; however, in real soils the electric field vector tilts and partly spreads into the earth, which dissipates power and transforms into heat [2]. This constitutes losses on earth.

Apparent soil electrical conductivity (ECa) to agriculture has its origin in the measurement of soil salinity, in arid-zone problem, which is associated with irrigated agricultural land. ECa is a quick, reliable, easy-to-take soil measurement that often relates to crop yield. For these reasons, the measurement of ECa is among the most frequently used tools in precision agriculture research for the spatiotemporal characterization of edaphic and anthropogenic properties that influence crop yield [37]. There are portable instruments for measuring the electrical conductivity of the soil by the method of electromagnetic induction and by the method of the four conductors, which are installed in the agricultural machinery to obtain a map of the soil, before carrying out the work of tilling the earth.

For geophysics applications, the solar disturbances (flares, coronal mass ejections) create variations of the Earth's magnetic field. These geomagnetic variations induce a geoelectric field at the Earth's surface and interior. The geoelectric field in turn drives geomagnetically induced currents, also called telluric currents along electrically conductive technological networks, such as power transmission lines, railways, and pipelines [38]. This geomagnetically induced currents create conditions where enhanced corrosion may occur. Earth conductivity can create geomagnetically induced current variations, in particular where a pipeline crosses a highly resistive intrusive rock. It is important to make pipeline surveys once a year to measure the voltage at test posts to ensure that pipe-to-soil potential variations are within the safe range, impressed by cathodic protection systems [38].

## Author details

Walter Gustavo Fano
Faculty of Engineering, Universidad de Buenos Aires, Buenos Aires, Argentina

*Address all correspondence to: gustavo.gf2005@gmail.com

## References

[1] Joffe E, Kai-Sang L. Grounds for Grounding: A Circuit to System Handbook. Piscataway, NJ, USA: Wiley, IEEE Press; 2009

[2] Trainotti V, Fano WG, Dorado L. Ingenieria Electromagnetica, Vol. 2. 1st ed. Buenos Aires, Argentina: Nueva Libreria; 2006

[3] Norton KA. The physical reality of space and surface waves in the radiation field of radio antennas. Proceedings of the Institute of Radio Engineers. 1937; **25**:1192-1202

[4] Angulo I, Barclay L, Chernov Y, Deminco N, Fernandez I, Gil U, et al. Handbook on Ground Wave Propagation. Geneva, Switzerland: Radiocommunication Bureau, ITU; 2014

[5] Corwin DL, Lesch SM. Application of soil electrical conductivity to precision agriculture. American Society of Agronomy. 2003;**95**(3):455-471

[6] Smith-Rose RL. The electrical properties of soil for alternating currents at radio frequencies. Proceedings of the Royal Society of London. 1933;**140**:359-377

[7] Jol HM. Ground Penetrating Radar Theory and Applications. Oxford, UK: Elsevier Science; 2009

[8] Ramo S, Whinnery J, Van Duzer T. Fields and Waves in Communication Electronics. India: Wiley; 2008

[9] Griffiths DJ. Introduction to Electrodynamics. 4th ed. Boston, MA: Pearson; 2013. Re-published by Cambridge University Press in 2017

[10] Reitz JR, Milford FJ, Christy RW. Foundations of Electromagnetic Theory. 4th ed. USA: Addison-Wesley Publishing Company; 2008

[11] Trainotti V, Fano WG. Ingenieria Electromagnetica, Vol. 1. 1st ed. Buenos Aires, Argentina: Nueva Libreria; 2004

[12] Stratton JA. Electromagnetic Theory. Hoboken, New Jersey, USA: IEEE Press, Wiley-Interscience; 2007

[13] Samoulian A, Cousin I, Tabbagh A, Bruand A, Richard G. Electrical resistivity survey in soil science: A review. Soil and Tillage Research. 2005; **83**(2):173-193

[14] Abu-Hassanein ZS, Benson CH, Blotz LR. Electrical resistivity of compacted clays. Journal of Geotechnical Engineering. 1996;**122**(5): 397-406

[15] Hoekstra P, Delaney A. Dielectric properties of soils at uhf and microwave frequencies. Journal of Geophysical Research. 1974;**79**:11, 1699-1708

[16] Peplinski NR, Ulaby FT, Dobson MC. Dielectric properties of soils in the 0.3–1.3-GHz range. IEEE Transactions on Geoscience and Remote Sensing. 1995;**33**:803-807

[17] El-rayes MA, Ulaby FT. Microwave dielectric spectrum of vegetation—Part I: Experimental observations. IEEE Transactions on Geoscience and Remote Sensing. 1987;**GE-25**:541-549

[18] Ulaby FT, El-rayes MA. Microwave dielectric spectrum of vegetation—Part II: Dual-dispersion model. IEEE Transactions on Geoscience and Remote Sensing. 1987;**GE-25**:550-557

[19] Hipp JE. Frequency dispersion of complex permeability in mnzn and nizn

spinel ferrites and their composite materials. Proceedings of the IEEE. 1974;**62**:98-103

[20] Fano WG, Trainotti V. Dielectric properties of soils. In: 2001 Annual Report Conference on Electrical Insulation and Dielectric Phenomena (Cat. No.01CH37225); 2001. pp. 75-78

[21] Karakash J. Transmission Lines and Filter Networks. New York, USA: Macmillan; 1950

[22] Peterson A, Durgin DG. Transient signals on transmission lines: An introduction to non-ideal effects and signal integrity issues in electrical systems. Synthesis Lectures on Computational Electromagnetics. 2008;**3**(1):1-144

[23] Collin R. Foundations for Microwave Engineering. 2nd ed. New York, USA: Wiley India Pvt. Limited; 2007

[24] Olchawa O, Kumor M. Time domain reflectometry (TDR) measuring dielectric constant of polluted soil to estimate diesel oil content. Archives of Hydro-Engineering and Environmental Mechanics. 2008;**55**(1–2):55-62

[25] Keysight. Time domain reflectometry theory. Keysight Technologies; Application Note; 2015

[26] Zante P. Water content and soil bulk electrical conductivity measurements. TDR Technology Applied to Soils. 2002;**49**(1):1-27

[27] Robinson DA, Jones SB, Wraith JM, Friedman SP. A review of advances in dielectric and electrical conductivity measurement in soils using time domain reflectometry. Vadose Zone Journal. Soil Science Society of America. 2003;**2**:444-475

[28] Evett SR. Coaxial multiplexer for time domain reflectometry measurement of soil water content and bulk electrical conductivity. Transactions of the ASAE. 1998;**41**(2): 361

[29] Ebrahimi-Birang N, Maul CP, Morley WA. Calibration of a TDR instrument for simultaneous measurements of soil water and soil electrical conductivity. Transactions of the ASABE (American Society of Agricultural and Biological Engineers). 2006;**49**(1):75-82

[30] Seyfried MS, Murdock MD. Measurement of soil water content with a 50-MHz soil dielectric sensor. Soil Science Society of America Journal. 2004;**68**:394-403

[31] Campbell JE. Dielectric Properties of Moist Soils at rf and Microwave Frequencies. New Hampshire, USA: Dartmouth Digital Library Collections; 1988

[32] ZChaudhari HC, Shinde VJ. Dielectric properties of soils at x-band microwave frequency. Indian Journal of Pure and Applied Physics. 2012;**50**(1): 64-66

[33] La Gioia A, Porter E, Merunka I, Shahzad A, Salahuddin S, OHalloran M. Open-ended coaxial probe technique for dielectric measurement of biological tissues: Challenges and common practices. Diagnostics. 2018;**8**(2):2-38

[34] Pozar D. Microwave Engineering. 4th ed. Hoboken, NJ, USA: Wiley; 2011

[35] Kirkscether EJ. Ground constant measurements using a section of balanced two-wire transmission line. IRE Transactions on Antennas and Propagation. 1960:307-312

[36] Fano W. Interacción de la radiación electromagnética Con Materiales dieléctricos y magnéticos. Aplicaciones. PHD Thesis. Universidad de Buenos Aires, Buenos Aires, Argentina; 2004

[37] Corwin D, Lesch S. Apparent soil electrical conductivity measurements in agriculture. Computers and Electronics in Agriculture. 2005;**46**(1):11-43

[38] Fernberg PA, Samson C, Boteler DH, Trichtchenko L, Larocca P. Earth conductivity structures and their effects on geomagnetic induction in pipelines. Annales Geophysicae. 2007; **25**(1):207-218

Chapter 4

# Thermoelectric Properties of Chalcogenide System

*Wiqar Hussain Shah and Waqas Muhammad Khan*


## Abstract

We will discuss the development of a new ternary and quaternary tellurium telluride chalcogenide nanoparticles used for efficient thermo-electric waste heat energy convertor called thermo-electric generator. Nanoparticles-based tellurium telluride chalcogenide nanoparticles, which will be used for thermoelectric generator, will eventually solve an important issue of the energy crises, that is, conversion of waste heat into useful electrical energy. By injecting charge carriers in the host matrix of $Tl_{10-x-y}A_xB_yTe_6$ nanomaterials system, different types of dopants (A = Pb, Sn, Ca and B = Pb, Sb Sr, etc.), with x = 0–2.5 and y = 0–2.5 on tellurium telluride has been introduced to synthesize new materials by Co-precipitation techniques and also by solid state reaction techniques followed by Ball-Milling for the fabrication of nanomaterials. We will study the effect of reduction of charge carriers in thermal and transport properties using different dopants contents by replacing host atoms. The charge carrier's concentration will affect the ratio of electron-hole concentration which in turns increases the electron scattering in these chalcogenide nanoparticles, which will affect the electrical conductivity and thermo-power. The prime purpose of doping with different ionic radii and different concentration is to enhance the power factor for the tellurium telluride nanosystem. At the end one will be able to control different physical parameters such as, thermally assisted electrical conductivity, and thermopower. Different characterization technique will be applied, for example, X-Ray diffraction techniques will be used for structural analysis, SEM will shows the morphological structure of the particles at 100 nm and energy dispersive x-rays spectroscopy will be used for elemental analysis. The electrical conductivity will be measured by four-probe resistivity measurement techniques, and Seebeck coefficient will be measured by standard temperature gradient techniques.


**Keywords:** effect of doping, Seebeck coefficient, electrical conductivity, power factor

## 1. Introduction

Energy storage and conversion devices (**Figure 1**) continue to be rich areas for scientific and engineering studies that incorporate novel features and functions in intelligent and interactive modes, represent a radical advance in consumer products, such as wearable electronics, healthcare devices, artificial intelligence, electric vehicles, smart household, and space satellites. However, there are still grand challenges in fundamental research and understanding to accelerate energy storage and

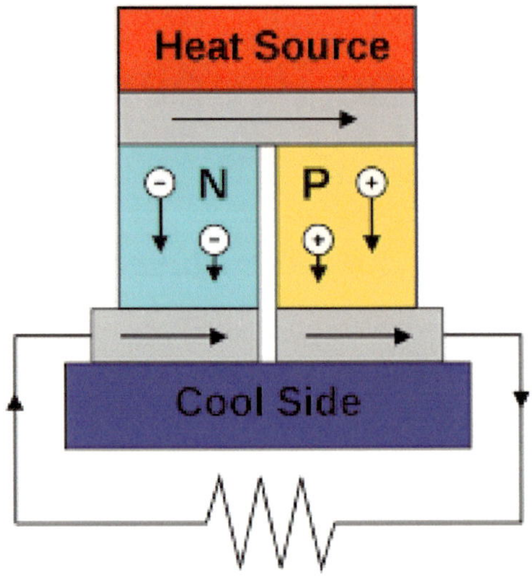


**Figure 1.**
*Energy storage and conversion devices.*

conversion devices to commercial reality, which include new materials and structures with high ionic conductivity, tailored mixed electron/ion conductivity, novel interface engineering methodologies, new device concepts, efficient and scalable techniques for materials and system-level integrations. This research study is intended to provide information for those working in energy storage and conversion devices from materials, characterizations, devices and system integrations to communicate recent progress on current technologies and to exchange ideas about next-generation solutions. In this research work, we will design and develop a new tellurium telluride chalcogenide materials used for efficient thermo-electric waste heat energy convertor called thermo-electric generator. Nanostructures based on tellurium telluride chalcogenide materials, which are used for efficient thermoelectric generator will eventually solve large issues of the energy crises.

Thomas john Seebeck discovered the effect of Seebeck on 14 December 1820, at famous science Academy of "Berlin" via detecting the deflection of magnetic compass needle nearby close ring of conducting wire that one adjacent is linked to metal of a low temperature while the second side is linked with metal of a high temperature. This effect recently verified that magnetic compass needle is bounced by reason of electrical current movement in wire and term "Thermo-magnetism" is changed by means of "the effect of thermoelectric". This electrical current/emf are similarly recognizing via means of "Seebeck emf." Thermo power "S" or Seebeck effect is the variation in electrical potential

$$(\Delta V = V_{hot} - V_{cold}) \tag{1}$$

Divided by the thermal gradient

$$(\Delta T = T_{\_hot} - T_{\_cold}), \tag{2}$$

Mathematically we can write it as,

$$S = \Delta V / \Delta T \tag{3}$$

$$S = (V_{\_cold} - V_{\_hot}) / (T_{\_hot} - T_{\_cold}) [V/K] \tag{4}$$

In 1834, The French watch maker and a part time physicist Jeane Chaarlese Athanaese Peltiere (1785–1845) published new research article in the French journal "Annal. Phy. Che," where Peltiere printed that by applying of electrical current in the direction of two dissimilar conductors, which are linked in series, disclosed that temperature changes by the side of joint of conductor. He recognized this specific effect as a "Peltier effect."

This particular effect can be mathematically written as,

$$Q = \Pi I \tag{5}$$

$$\Pi = \frac{Q}{I}\ [Watt/Ampere] \tag{6}$$

In upper formulation, the term "Q" representing degree by which warmth remains flow away per unit of time, while "$\Pi$" in formula is the "Peltiere coefficient" while the term "$I$" in upper formulation is electrical current which is formed in thermoelectric material (TEM).

Wiliam Thomoson in 1851 presented his well-known effect, which is named Thomson effect. "By means of creating a change in temperature "$\Delta T$" crossways a conductor, wherein he displays association among Seebeck and Peltier effect."

Mathematically it can be written as,

$$\frac{dq}{dt} = -KJ\Delta T \tag{7}$$

In the above equation "$K$" is Thomson's coefficient and "$J$" is the density of current.

The Seebeck, The Peltiere and The Thomsen's coefficients are connected through the equation,

$$\Pi = TS \tag{8}$$

### 1.1 The Seebeck Coefficient

Once material is being heated through one side, then due to thermal gradient the thermoelectric voltage rises, since charge carriers (holes/electrons) drift from hot to ward cold. An induced voltage amount over a gradient of temperature

$$\left(S = \frac{\Delta V}{\Delta T}\right) \tag{9}$$

Among materials two (2) edges is recognized as per coefficient of Seebeck. Its unit is volt per kelvin (v/k).

When taking a Seebeck coefficient into its constituent, then written as below,

$$S = T\frac{8\pi^2 k_B^2 m^*}{3eh^2}\left(\frac{\pi}{3n}\right)^{2/3} \tag{10}$$

The above equation is Seebeck coefficient of degenerate semiconductors and metals. Above equation comprises of three (3) variables: Variable "T" signifies temperature, variable "n" signifies charge carrier concentration, besides variable "$m^*$" signifies "effective mass"

## 1.2 The electrical conductivity

Charge movement in matter is named conductivity, it is represented by "$\sigma$". The movement of holes or free electrons in a particular way reasons an electrical current in matter. This electrical current in matter which in a particular way, is the consequence of movement of charges via smearing the potential change crossways a semiconductor or conductor. The equation for current density "$J$" is;

$$J = nq\nu_d \tag{11}$$

In above equation, "$n$" signifies carrier concentration, the q in equation signifies charge carrier which is equivalents for holes to $+1.602 \times 10^{-19} C$ while $-1.602 \times 10^{-19} C$ for e's, and "$\nu_d$" is called electron drift velocity. By means of putting value of drift velocity, $\nu_d = \frac{Ee\tau}{m^*}$ in Eq. (14), we obtained the relation which is used in the existence of constant electric field for current density, and this is as well-known as law of ohm.

$$J = \frac{ne^2\tau E}{m^*} \tag{12}$$

In terms of electrical conductivity and electric field, the current density $J$ is,

$$J = \sigma E \tag{13}$$

$$\sigma = \frac{ne^2\tau}{m^*} = ne\mu \tag{14}$$

Where

$$\mu = \frac{e\tau}{m^*} \tag{15}$$

While "$\mu$" is charge transporter movement and having dimension of $\left[\frac{cm^2}{Vs}\right]$. "$\tau$" is the scattering time while"$m^*$" is electron effective mass. The scattering time "$\tau$" can be described as, "it is the amount of time in which the charge transporter their momentum is places besides turn out to be in balance after the elimination of exterior electric field. The scattering time takes near relative through the electronegativity of an element.

## 1.3 Power factor

Through study of power factor (PF), the success of a cooler of thermoelectric cooler (TEC) besides generator (TEG) is resolute through study of power factor; it is represented via "PF," and calculated through Seebeck coefficient square multiplied through the electrical conductivity at precise temperature.

Mathematically it can be written as,

$$PF = S^2\sigma \tag{16}$$

In the above equation, $PF$ is Power Factor, $S$ is Seebeck coefficient while $\sigma$ is Electrical conductivity.

Thermoelectric devices having values of Seebeck coefficient (S) and electrical conductivities high, gives high power factor (PF) and too charitable high electrical power.

### 1.4 Objectives

Briefly, the specific objectives are as follows:

1. Fabrication of chalcogenides based materials via cost effective chemical synthesis to obtain nanomaterials. Specifically:

    a. Nanostructured n-type $Tl_5Te3$ via solution co-precipitation and thermos-chemical treatment.

    b. Nanostructured p-type Tl5Te3 via solution co-precipitation and fast chemical reduction.

2. Optimization of critical SPS parameters (such as sintering temperature, applied pressure, holding time and heating rates) for chalcogenides while consolidating these materials to preserve the nanostructure, to reduce thermal conductivity.

3. Bottom-up chemical synthesis and detailed characterization for low temperature TE applications.

4. Fabrication of silicide based TE materials through mechanical alloying (top-down approach). Specifically: a. n-type Tl5Te3 by ball milling for an optimized reaction time and followed by materials' characterizations to identify the phase of the materials.

    a. Doping of Sb and Pb in n-type Tl5Te3 nanomaterials and to investigate its effect on TE performance.

    b. Fabrication of p-type HMS via ball milling by utilizing optimized react followed by detailed physiochemical characterizations.

    c. Study the effect of ternary and quaternary tellurium telluride chalcogenide nanoparticles as nanoinclusions/grain boundary.

5. Optimization of SPS critical parameters (such as sintering temperature, applied pressure, holding time and heating rates) while consolidating these materials to preserve the nanostructure and obtain the desired phases.

### 1.5 Electrical conductivity and Seebeck coefficient measurements

**Figure 2** shows the measurement system that was used to measure the electrical properties of the samples. **Figure 2(a)** displays the photograph of the commercial ZEM-3 system and **Figure 2(b)** shows a sample mounted in the ZEM-3 apparatus for electrical resistivity and Seebeck coefficient measurement. In the ZEM-3 system, electrical resistivity was measured using a four probe technique and electrical conductivity was calculated from the electrical resistivity. The four probe technique for measuring the resistivity simply accounts for the contact resistance between metal electrodes and the semiconducting samples. **Figure 3(a)** displays a schematic diagram of the four probe used by the ZEM-3 system. As shown in **Figure 3(a)**, in the four probe technique current I was passed through one set of probes (blue blocks) and the voltage difference ($\Delta V$) was measured using another set of probes (small red spheres). These four probes were connected to four thermocouples. The voltage and

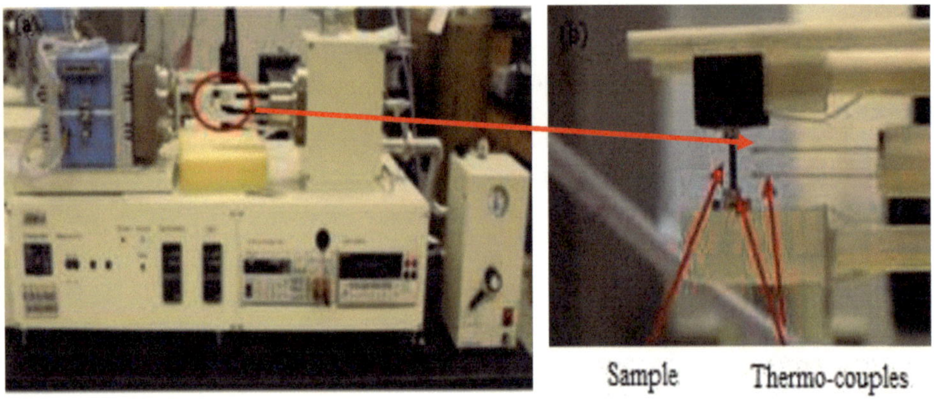


**Figure 2.**
*(a) A commercial ZEM-3 system and (b) magnified sample holder region [indicated by red circle in (a)] with a sample mounted for measurement.*

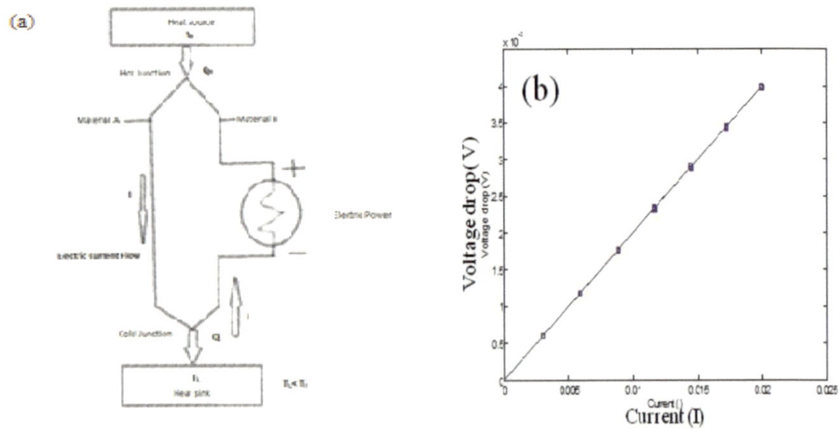


**Figure 3.**
*(a) Displays a schematic diagram of the four probe used by the ZEM-3 system and (b) a typical I-V curve for resistance measurement.*

current control, data acquisition, and interpretation were fully automated and computer controlled. The electrical resistivity was found from the relation,

$$\rho = \frac{A}{l}\left(\frac{\Delta v}{\Delta I}\right) \tag{17}$$

Where $(\Delta V/\Delta I)$ is the slope of the I-V curve as shown in **Figure 3(b)**, $A$ is the cross-sectional area of the sample and $l$ is the distance between the voltage probes.

The electrical conductivity was then calculated as the reciprocal of the resistivity.

$$\sigma \, \alpha \frac{1}{\rho} \tag{18}$$

During the resistivity measurement, the temperatures at both probes were kept constant to minimize the Seebeck voltage. The same ZEM-3 system (**Figure 2**) was used for Seebeck coefficient measurement.

The Seebeck coefficient is simply defined as the ratio of an open-circuit potential difference ($\Delta V$) to a temperature gradient ($\Delta T$),

$$S = \frac{\Delta V}{\Delta T} \tag{19}$$

For Seebeck coefficient measurement, the voltage and temperatures were measured simultaneously by the same thermocouple probe (small red spheres) as shown in **Figure 3(a)**. Then, the voltage difference ($\Delta V$) was measured for a set of temperature differences ($\Delta T$) between the two probes and the Seebeck coefficient was calculated from the slope of $\Delta V$- $\Delta T$ plot.

Thallium antimony telluride $TlSbTe_2$ nanoparticles have been prepared by co-precipitation techniques. We have investigated that the electrical resistivity is high and the thermal conductivity is low as compared to Sintered $Bi_2Te_3$ and TAGS "$(GeTe_{1\text{-}x}(AgSbTe_2)_x$" material. The Seebeck coefficient of $TlSbTe_2$ is 224 $\mu vk^{-1}$ at 666 k which is positive in the whole temperature range showing p-type behavior. The power factor ($S^2\sigma$) found for $TlSbTe_2$ is $8.9 \times 10^{-4}$ $Wm^{-1}k^{-2}$ at 576 K which is low as compared the power factor of current thermoelectric devices, that is, in the range $10^{-3}$ $Wm^{-1}k^{-2}$. The figure of merit (ZT) of the order of 0.87 was found at 715 k for $TlSbTe_2$ [1].

Prepared a new low-valent thallium silicon telluride $Tl_6Si_2Te_6$ and compared there results crystal and electronic structure and there electronic properties with $Tl_6Ge_2Te_6$, they observed the same crystal structure of $Tl_6Si_2Te_6$ with $Tl_6Ge_2Te_6$, the quantitative results for $Tl_6Si_2Te_6$. The demerit $[\boldsymbol{Si_6Te_6}]^{2-}$ units crystal structure was found with a single Si-Si bond, the weak bond exist among Tl-Tl and irregularly coordinated by 5 or 6 Te atoms, the black color was observed for both compounds exhibiting a small band gap of the order **0.9 ev** for $Tl_6Si_2Te_6$ and **0.5 ev** for $Tl_6Ge_2Te_6$ compounds. The electrical conductivity and Seebeck coefficient investigated for $Tl_6Si_2Te_6$ is $\mathbf{5.5\ \Omega^{-1}cm^{-1}}$ and $+\mathbf{65\ \mu vK^{-1}}(\boldsymbol{at\ 300\ K})$, while for $Tl_6Gd_2Te_6$ is $\mathbf{3\ \Omega^{-1}cm^{-1}}$ and $+\mathbf{150\ \mu vK^{-1}}(\boldsymbol{at\ 300\ K})$ [2].

Prepared the samples of polycrystalline $Ag_9TlTe_5$, the different nominal composition by them are: $Ag_9TlTe_x(X\ 5.0, 5.05, 5.1, 5.2, 5.3, 5.5, 5.7, 6.0)$, the $Ag_9TlTe_x$ samples were made by heating the $Ag_2Te$, $Tl_2Te$, $Tl_2Te_{1.2}$ and Te with a proper quantity in sealed quartz tube, the ball-milling and hot-press techniques were used for construction of required samples. The X-Ray Diffraction (XRD) technique were used for analysis of phase relation, the electrical resistivity is calculated which is decreasing with increasing temperature except for $Ag_9TlTe_{5.0}$, the investigated Seebeck coefficient is almost independent of temperature except for $Ag_9TlTe_{5.0}$, the power factor is quite high of the order of $0.3 \sim 0.4 \times 10^{-3}$ $Wm^{-1}K^{-2}$ while the power factor for $Ag_9TlTe_{5.0}$ is quite low of the order of$0.05 \times 10^{-3}$ $Wm^{-1}K^{-2}$. The dimensionless thermoelectric figure of merit investigated for $Ag_9TlTe_{5.0}$ is very low approximately 0.08 while for $X \geq 5.0$ is high approximately 1.0, which shows that all the physical properties are changing by changing tellurium content in $Ag_9TlTe_x$ [3].

Herman et al. [4] uses the concept of electronic density of states to increase the thermoelectric figure of merit in lead telluride PbTe, the Seebeck coefficient was enhanced by deforming the electronic density of states, leads to double the thermoelectric figure of merit and he further explained that in nanostructured material it may give us further good results [4].

Synthesized $Tl_4MTe_4$ (M = Zr, Hf) and investigated their crystal structure and thermoelectric properties, for investigating their crystal structure, the X-ray diffraction technique were used, they found that the crystal structure of $Tl_4MTe_4$ is octahedral with a space group $R\bar{3}$, the unit cell dimension for $Tl_4ZrTe_4$ is

$a = 14.6000(5)$ Å and $c = 14.189(1)$ Å, and $Tl_4HfTe_4$ the unit cell dimension is $a = 14.594(1)$ Å and $c = 14.142(3)$ Å. Linear muffin-tin orbital (LMTO) methods were used for the calculation of electronic structure clearing that $Tl_4MTe_4$ exhibit semiconducting behavior exhibiting an indirect band gap of0.3 ev. They observed that the electrical resistivity and Seebeck coefficient decreased while the thermal conductivity increased temperature, the thermoelectric figure of merit (ZT) for $Tl_4ZrTe_4$ compound increased from 0.14 to 0.1 between the temperature 370 K and 420 K while decreasing when the temperature increased from 420 K, for $Tl_4HfTe_4$ the ZT varies from 0.05 to 0.09 between temperature 370 K and 540 K [5].

In the other research, prepared the ternary compound $Tl_2ZrTe_3$ and compared it with $Tl_2SnTe_3$ and investigated the different properties such as structural, physical and thermal properties. $Tl_2ZrTe_3$ compound exhibits a simple cubic structure with a lattice parameter $a = 19.118(1)$ Å $(Z = 36)$. They investigated the electronic properties which clears that the $Tl_2ZrTe_3$ exhibits semiconducting behavior. The band gap observed for $Tl_2ZrTe_3$ is 0.7 ev which is higher from the band gap of $Tl_2SnTe_3$ ($Eg = 0.4$ ev), the electrical conductivity is independent of temperature ranging from room temperature to 450 K, when the temperature rises from 450 K the electrical conductivity falls abruptly, while for $Tl_2SnTe_3$ compound the electrical conductivity decreases from 22 $\Omega^{-1}cm^{-1}$ to 15 $\Omega^{-1}cm^{-1}$ with increasing temperature from room temperature to 515 K, from 373 K to 450 K the thermal conductivity decreases from 0.39 $Wm^{-1}K^{-1}$ to 0.30 $Wm^{-1}K^{-1}$ for $Tl_2ZrTe_3$ and for $Tl_2SnTe_3$ the thermal conductivity decreases from 0.24 $Wm^{-1}K^{-1}$ at 420 K to 0.20 $Wm^{-1}K^{-1}$ at 450 K. For $Tl_2ZrTe_3$ the calculated Seebeck coefficient is generally same in the range of 373 K to450 K, but decreases with temperature greater than450 K, the peak experiential value of Seebeck coefficient is120 $\mu VK^{-1}$. For $Tl_2SnTe_3$ the Seebeck coefficient increases from 240 $\mu VK^{-1}$ to 330 $\mu VK^{-1}$ with increasing temperature from room temperature to 450 K. The calculated power factor for $Tl_2ZrTe_3$ is almost same changes from 0.35 $\mu Wcm^{-1}K^{-1}$ to 0.41 $\mu Wcm^{-1}K^{-1}$ while for $Tl_2SnTe_2$ the power factor is increasing with temperature. The ZT value of 0.18 at 450 K was observed for $Tl_2ZrTe_3$ which is less than ZT value of $Tl_2SnTe_3$ (ZT = 0.31 at 500 K) [6].

Studied the thermoelectric properties of ternary thallium chalcogenides $TlGdQ_2$ (Q = SE, Te) and $Tl_9GdTe_6$. They found that $TlGdQ_2$ is isostructural with $TlSbQ_2$ and $Tl_9GdTe_6$ is isostructural with $Tl_9BiTe_6$. They found the high Seebeck coefficient and low electrical conductivity of $TlGdQ_2$. The low thermal conductivity of the order of 0.5 $Wm^{-1}K^{-1}$was investigated at room temperature for $TlGdTe_2$. In the studies of $Tl_9GdTe_6$ they found the low power factor due to the high electrical conductivity of 850 $\Omega^{-1}cm^{-1}$ and low seebeck coefficient of 27 $\mu VK^{-1}$ at550 K. In the whole study they found the good thermoelectric property for $TlGdTe_2$, the dimensionless figure of merit is 0.5 at 500 K [7].

Prepared the thallium lanthanide telluride $Tl_{10-x}Ln_xTe_6$ with Ln = CE, Pr, Nd, Sm, Gd, Tb, Dy, Ho and Er and $0.25 \le x \le 1.32$ by hot press method. He found that the crystal structure of $Tl_{10-x}Ln_xTe_6$ is isostructural to $Tl_9BiTe_6$ and the volume of unit cell is increases with increasing lanthanide content. They investigated that the electrical and thermal conductivity decreases with increasing the lanthanide content while the Seebeck coefficient is increases. In this series for **$Tl_{8.97}Ce_{1.03}$ $Te_6$, $Tl_{8.92}Pr_{1.08}Te_6$ and $Tl_{8.99}Sm_{1.01}Te_6$** the electrical and thermal conductivity increases and Seebeck coefficient value decreases due to the discontinuity in the band gap as compared to $Tl_9LnTe_6$ compounds. The power factor is increases with increasing the lanthanide content "x" in the $Tl_{9-x}Ln_xTe_6$ resulting the increase in the dimensionless thermoelectric figure of merit, the best thermoelectric figure of merit values are 0.22 at 550 K were achieved for the stoichiometric compound on cold press pellets [8].

Fabricate and improve the thermoelectric properties of alumina nanoparticle-dispersed $Bi_{0.5}Sb_{1.5}Te_3$ matrix composites, the nanoparticles were fabricated by ball milling process and followed by spark plasma sintering process. The p-type bismuth antimony telluride (BST) nanopowder prepared from the mechano-chemical process, were mixed with 1.0, 0.5, and 0.3 vol.% with $Al_2O_3$ nanoparticles by ball milling process. They studied the surface morphology and investigated the size of the nanoparticles. The electrical resistivity is increasing with temperature from $1.5 \times 10^{-5}$ Ωm at 293 K to $2.5 \times 10^{-5}$ Ωm at 473 K, and the seebeck coefficient is increases from $+205\ \mu VK^{-1}$ to $+210\ \mu VK^{-1}$ from temperature 293 K to 473 K, respectively, shows p-type semiconducting behavior. The highest Seebeck coefficient of the order of 235 $\mu VK^{-1}$ at 373 K was observed from 0.3vol.% $Al_2O_3$/BST nanocomposite. They show that the increasing volume fraction of $Al_2O_3$ increases the carrier density which affects the Seebeck coefficient, carrier mobility and electrical resistivity of $Al_2O_3$/BST nanocomposites. The observed power factor is 1.7 times higher from pure BST, that is, 33 $\mu WK^{-2}$cm at 393 K for 0.3vol.% $Al_2O_3$/BST nanocomposites, and for pure BST is22 $\mu WK^{-2}$cm. The thermal conductivity is decreased by the addition of $Al_2O_3$ nanoparticles, for pure BST is 0.8 $Wm^{-1}K^{-1}$ and for $Al_2O_3$/BST is0.7 $Wm^{-1}K^{-1}$. The thermoelectric figure of merit (ZT) is 1.5 for 0.3vol.% $Al_2O_3$/BST composite at 373 K which is higher from pure BST [9].

Studied the thermoelectric properties of Indium doped SnTe ($In_xSn_{x-1}Te$) nanostructured compound. They prepared $In_xSn_{x-1}Te$ by ball milling and hot press techniques, and investigated their thermal conductivity, diffusivity and electrical conductivity decreases with temperature ranging from 300 to 900 K, while the power factor and Seebeck coefficient is increases and the specific heat is almost constant a little increase was seen in $In_xSn_{x-1}Te$. The sample is also prepared by ball mill and hand mill method but the result is the same. They observed the relationship of carrier concentration vs. Seebeck effect which shows that In doped SnTe shows abnormal behavior with increasing carrier concentration, they get the SEM, TEM, and HRTEM images of $In_xSn_{x-1}Te$ which clearly shows the sample is consist of both small and large grain boundaries with a good crystallinity which effects the thermal conductivity of the sample. The thermal electric figure of merit is observed which greater than 1 is at temperature 873 K in for $In_{0.0025}Sn_{0.9975}Te$ [10].

Investigated the thermoelectric properties of indium doped $PbTe_{1-y}Se_y$ alloys, solid state method was used for the synthesis. They showed that the carrier concentration and electrical resistivity increases with increasing temperature in n-type indium doped $PbTe_{1-y}Se_y$ alloys affecting the Seebeck coefficient and as a result the power factor of $PbTe_{1-y}Se_y$ alloys. The bipolar effect was observed at high temperature which restricts the thermoelectric figure of merit to 0.66 at 800 K with 30% content in sample; they further concluded that for the enhancement of thermoelectric properties, the increased carrier concentration must be reduced at high temperature [11].

Optimized the thermoelectric properties of $Tl_{10-x-y}Sn_xBi_yTe_6$, a quaternary telluride series has been studied. The crystal structure was investigated by X-Ray diffraction which belongs to $Tl_5Te_3$ type structure and the volume is increases with increasing the Sn concentration in $Tl_{10-x-y}Sn_xBi_yTe_6$, the electronic structure calculation revealed that $Tl_{8.5}SnBi_{0.5}Te_6$ is a narrow band gap p-type intrinsic semiconductor and $Tl_9Sn_{0.5}\ Bi_{0.5}Te_6$ is a p-type and narrow band gap extrinsic semiconductor. The electrical conductivity is decreasing with increasing temperature for $Tl_9Sn_{1-y}Bi_yTe_6$, $Tl_{8.67}Sn_{1-y}\ Bi_yTe_6$ and $Tl_{8.33}Sn_{1.12}Bi_{55}Te_6$. The low increase of the order of 0.4 $Wm^{-1}K^{-1}$ to 1.4 $Wm^{-1}K^{-1}$was observed in the thermal conductivity of

$Tl_{10-x-y}Sn_xBi_yTe_6$ type materials due to thermal conductivity of electron $\kappa_{el}$, and for $Tl_9Sn_{0.2}Bi_{0.8}Te_6$ the thermal conductivity decreases with increasing temperature due to the increase in lattice vibration. The thermoelectric figure of merit, Seebeck coefficient and power factor was increased with increasing temperature, The high power factor $S^2\sigma = 8.1\ \mu Wcm^{-1}K^{-2}$ was observed for $Tl_9(Sn, Bi)Te_6$ type system, they identified that with low Sn concentration the Seebeck coefficient and power factor is high and at low temperature the power factor is decreases. Yi et al. were study that bulk crystalline ingots into nanopowders by ball milling and hot pressing, in nanostructured bulk bismuth antimony telluride had achieved high figure of merit. He obtained a high value of ZT$\sim$ 1.3 in the temperature range of 75 and 100°*C*. The improvement of ZT is mostly due by the lower of thermal conductivity. TEM observation of microstructure show that the lower thermal conductivity due to the increased of phonon scattering, and increasing of phonon scattering due to the increasing of grain boundaries of nanograins, nanodots, precipitates and defects. When ingot was used as the starting material, the highest ZT value of $\sim$ 0.7 was obtained for bismuth antimony telluride [12].

It is very important for the control of carrier concentration, which is good for thermoelectric properties. Another approach like nanostructuring, the dimensionless figure-of-merit is increased by the reduction of lattice thermal conductivity.

## 2. Experiment

For the preparation of $Tl_{10-x-y}A_xB_yTe_6$ compounds of different types of dopants (x = Pb, Sn, Ca and y = Pb, Sb, Sr, etc.), with different concentration on tellurium telluride has been prepared by solid state reactions in evacuated sealed silica tubes. The purpose of this study were mainly for discovering new type of ternary and quaternary compounds by using $Tl^{+1}$, $Sn^{+3}$,$Pb^{+3}$ and etc. $Te^{-2}$ elements as the starting materials [13]. Direct synthesis of stoichiometric amount of high purity elements, that is, 99.99% of different compositions have been prepared for a preliminary investigation. Since most of these starting materials for solid state reactions are sensitive to oxygen and moistures, they were weighing stoichiometric reactants and transferring to the silica tubes in the glove box which is filled with Argon. Then, all constituents were sealed in a quartz tube. Before putting these samples in the resistance furnace for the heating, the silica tubes was put in vacuum line to evacuate the argon and then sealed it. This sealed power were heated up to 650 C$^o$ at a rate not exceeding 1 k/mint and kept at that temperature for 24 hours. The sample was cooled down with extremely slow rate to avoid quenching, dislocations, and crystals deformation. The nanoparticles have been prepared by ball-milling techniques.

Structural analysis of all these samples was carried out by x-rays diffraction, using an Intel powder diffractometer with position-sensitive detector and CuKα radiation at room temperature. No additional peaks were detected in any of the sample discussed here. X-ray powder diffraction patterns confirm the single phase composition of the compounds.

The temperature dependence of Seebeck coefficient was measured for all these compounds on a cold pressed pellet in rectangular shape, of approximately $5 \times 1 \times 1$ mm$^3$ dimensions. The air sensitivity of these samples was checked (for one sample) by measuring the thermoelectric power and confirmed that these samples are not sensitive to air. This sample exposes to air more than a week, but no appreciable changes observed in the Seebeck values. The pellet for these measurements was annealed at 400 C$^0$ for 6 hours. For the electrical transport measurements (electrical conductivity) four-probe resistivity technique was used and the pellets were cut into rectangular shape with approximate dimension of $5 \times 1 \times 1$ mm$^3$.

## 3. Results and discussions

### 3.1 Ternary system

#### *3.1.1 Structural analysis*

Various concentrations of *Sn* doped $Tl_{10-x}Sn_xTe_6$ compounds series were synthesized, and their physical properties were studied for x = 1, 1.25, 1.50, 1.75, 2.00. The powder x-rays diffraction pattern which is measured at room temperature for all these compounds is presented in **Figure 4**. It is found that the tetragonal single phase $Tl_8Sn_2Te_6$ is obtained in the present study. The tetragonal lattice parameters observed at room temperature are *a* = 8.8484 nm, and c = 13.0625 nm in table. The materials are iso-structural with the binary tellurium telluride $Tl_{10}Te_6$, and the crystal structure of $Tl_9Sb_1Te_6$ was determined with the experimental formula, possessing the same space group 14/mcm as $Tl_5Te_3$ and $Tl_8Sb_2Te_6$, in contrast to $Tl_9Sb_1Te_6$ that adopt the space group 14/m. **Figure 5** shows the SEM and EDX images. The SEM shows the morphology of the ternary compound at 100 nm scale.

### 3.2 Physical properties

#### *3.2.1 Electrical conductivity measurements*

We were interested in the effects of Sn doping in parent composition on the border line of semi-conductor and metallic. The experiment was conducted in a commercial oxford instrument cryostat with temperature control better than 0.5 $K$. the contacts were standard 4-probe, and were made using high quality silver point, contacts resistance was checked at room temperature and experiments were only carried out if it was satisfactorily low, typical current used were of the rate of 0.1 $mA$.

The electrical properties of tin doped thallium telluride nanostructural system has been investigated under dependency of temperature varying from room temperature 300 to 650$K$ by four probe resistivity technique. It has been founded that $Tl_8Sn_2Te6$ has the lowest electrical conductivity of $471.6\Omega^{-1}cm^{-1}$, and when Sn dopant increases from $x = 1.0$ to $x = 2.0$, an electrical conductivity decreases, for example, $\sigma = 471.68\ \Omega^{-1}cm^{-1}$ at $x = 2.0(20\%)$ and $\sigma = 1629.21\Omega^{-1}cm^{-1}$ at $x = 1.0(10\%)$ as shown in **Figure 6**. The electrical conductivity, σ, for the $Tl_{10-x}Sn_xTe_6$ samples with $x < 2.0$ decreases with increasing temperature, **Figure 6**, which clarifies that this decrease in the electrical conductivity is due to the high charge carrier

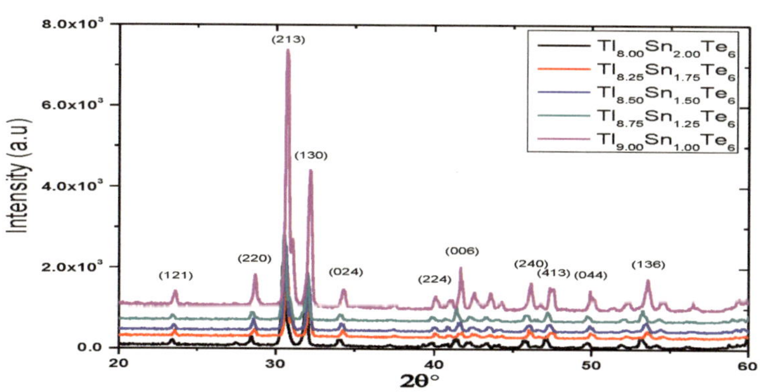


**Figure 4.**
*X-ray diffractometery of $Tl_{10-x}Sn_xTe_6$ doping Sn = 1, 1.25, 1.50, 1.75 and 2.*

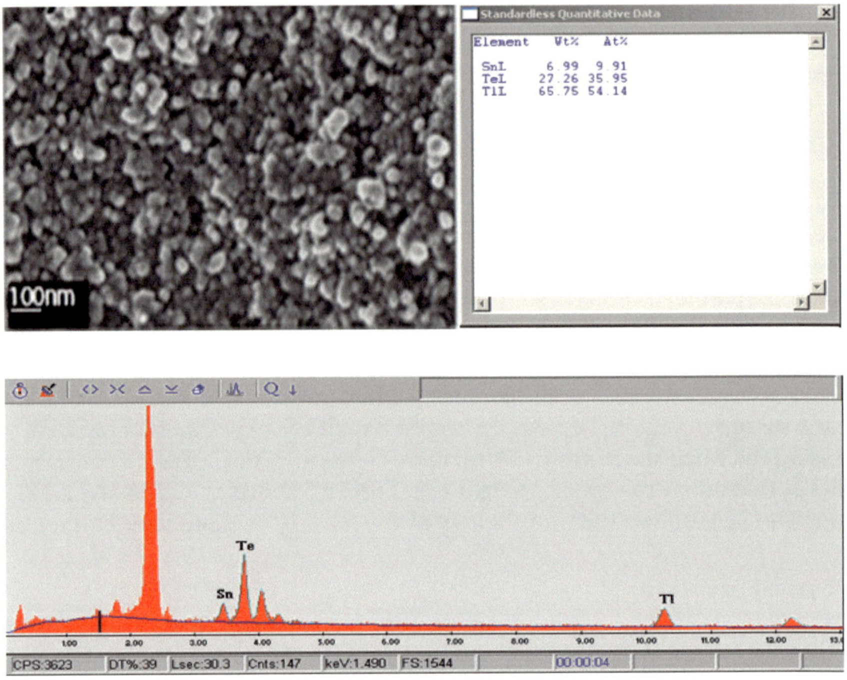


**Figure 5.**
*SEM and EDX image of $Tl_8Sn_2Te_6$.*

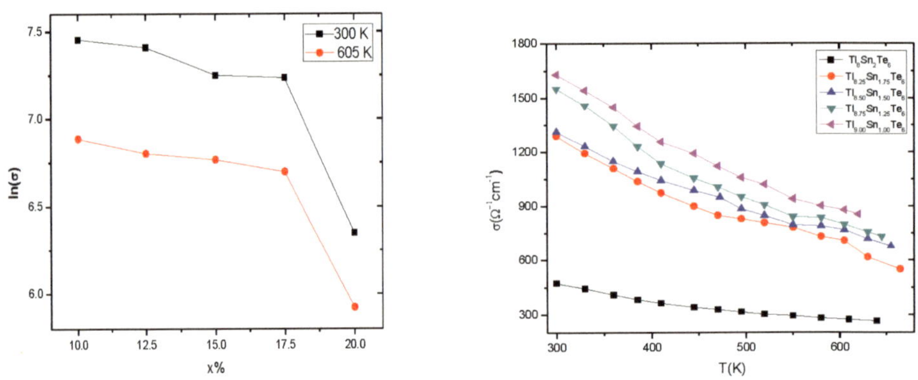


**Figure 6.**
*Electrical conductivity measurements at different concentration and high temperature.*

concentration. The highest value has been observed for $Tl_9Sn_1Te_6$ and the lowest value has been observed in sample $Tl_{8.0}Sn_{2.0}Te_6$ at $300K$. **Figure 6**, shows that the increases the concentration of the dopant decreases the electrical conductivity with temperature of the ternary compounds.

### *3.2.2 Seebeck coefficient measurements*

The positive Seebeck coefficient, S, observed in all samples of $Tl_{10\text{-}x}Sn_xTe_6$ as shown in **Table 1**, which increases smoothly with increasing temperature with $1.0 \le x \le 2.0$, for p-type semiconductors having high charge carrier concentration (**Table 2**). The Seebeck curve of the sample with x = 2.0 exhibits a clear

| Sample | Crystallite size, $D = 0.9\lambda/\beta\cos\theta$ (nm) | Lattice constant a, b, c = (Å) | Volume (Å$^3$) |
|---|---|---|---|
| $Tl_9Sn_1Te_6$ | 62.919 | a = b = 8.7930<br>c = 13.0050 | 1005.505 |
| $Tl_{8.75}Sn_{1.25}Te_6$ | 63.965 | a = b = 8.8450<br>c = 13.0755 | 1022.948 |
| $Tl_{8.50}Sn_{1.50}Te_6$ | 66.2833 | a = b = 8.8250<br>c = 13.0000 | 1012.44 |
| $Tl_{8.25}Sn_{1.75}Te_6$ | 59.820 | a = b = 8.8100<br>c = 13.0010 | 1009.086 |
| $Tl_8Sn_2Te_6$ | 56.793 | a = b = 8.8484<br>c = 13.0625 | 1022.717 |

**Table 1.**
*Crystallite size, lattice constant, and volume of unit cell.*

| Sample | Electrical conductivity ($\Omega^{-1}cm^{-1}$) at 300K | Electrical conductivity ($\Omega^{-1}cm^{-1}$) at 650K |
|---|---|---|
| $Tl_9Sn_1Te_6$ | 1629.21 | 939.137 |
| $Tl_{8.75}Sn_{1.25}Te_6$ | 1550.997 | 841.481 |
| $Tl_{8.50}Sn_{1.50}Te_6$ | 1310.326 | 795.66 |
| $Tl_{8.25}Sn_{1.75}Te_6$ | 1287.64 | 781.313 |
| $Tl_8Sn_2Te_6$ | 471.68 | 292.102 |

**Table 2.**
*Electrical conductivity of $Tl_{10\text{-}x}Sn_xTe_6(1 \leq x \leq 2)$ at 300 K and 650 K.*

maximum of $S = 79.77 + \mu V.K^{-1}$ at $300K$ and $S = +157.931\mu V.K^{-1}$ at $550\ K$. The lowest Seebeck coefficient $S = +33.15\ \mu VK^{-1}$ at $300K$ has been observed for $x = 1.0$ which is increases to$S = +65.84\mu V.K^{-1}$ at 550 K, it has been declared that when Sn content increase in host sample the Seebeck coefficient $S$, also increases, for example, for $x = 1.0$ the Seebeck coefficient has been observed is $S = +33.15\mu VK^{-1}$ to $S = +79.77\mu V.K^{-1}$ for at $300\ K$ and the Seebeck coefficient has been observed is $S = +65.844\ \mu VK^{-1}$ to $S = +157.937\mu V.K^{-1}$ for $x = 2.0$ at $550\ K$, respectively as shown in **Figure 7**. The power factor of $S_n$ doping is increases as the temperature is increase as shown in **Figure 8**.

### *3.2.3 Power factor analysis*

The power factor investigations show that it increases because of increasing behavior in Seebeck coefficient with temperature (**Table 3**). The calculated power factor is directly proportional to the square of the Seebeck coefficient and the electrical conductivity. The lowest power factor $1.9\mu Wcm^{-2}K^{-2}$ has been observed for $Sn = 1$ and the highest power factor $7.579\mu Wcm^{-2}K^{-2}$ has been observed for $Sn = 1.75$ at temperature $300K$, while for $Sn = 2$ the power factor is $3.639\mu Wcm^{-2}K^{-2}$ as shown in **Table 4**, which is increases with increasing temperature; this low power factor has been observed due to extremely low electrical conductivity of $Tl_8Sn_2Te_6$ which play a man role in the investigation of power factor.

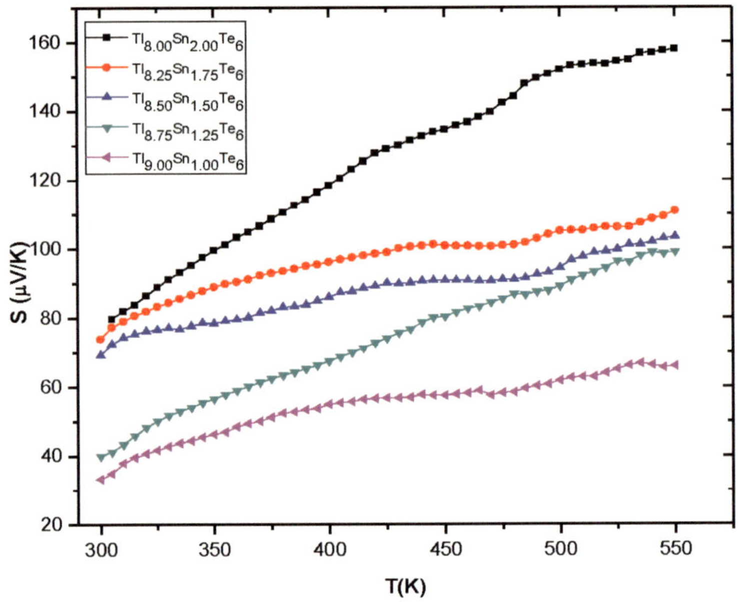


**Figure 7.**
*Seebeck coefficient measurements of $Tl_{10x}Sn_xTe_6$ $(1 \leq x \leq 2)$ at 300K and 550K.*

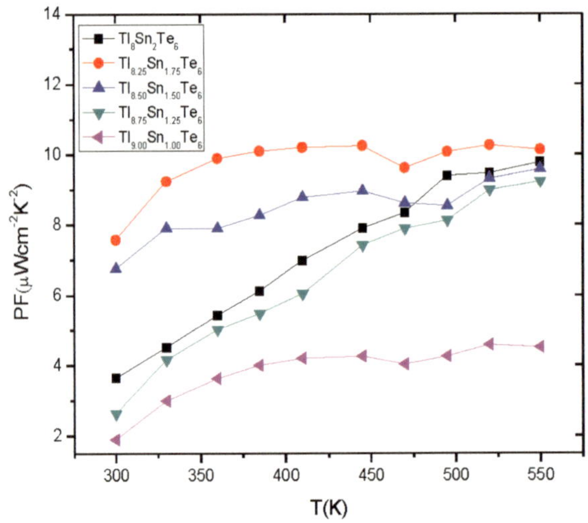


**Figure 8.**
*The power factor (PF) of $Tl_{10-x}Sn_xTe_6$ with $x = 1, 1.25, 1.50, 1.75,$ and 2.*

| Sample | Seebeck coefficient ($\mu VK^{-1}$) at 300K | Seebeck coefficient ($\mu VK^{-1}$) at 550K |
|---|---|---|
| $Tl_9Sn_1Te_6$ | 33.15 | 65.844 |
| $Tl_{8.75}Sn_{1.25}Te_6$ | 39.953 | 99.035 |
| $Tl_{8.50}Sn_{1.50}Te_6$ | 69.207 | 103.419 |
| $Tl_{8.25}Sn_{1.75}Te_6$ | 73.879 | 110.958 |
| $Tl_8Sn_2Te_6$ | 79.77 | 157.931 |

**Table 3.**
*Thermoelectric properties of $Tl_{10-x}Sn_xTe_6(1 \leq x \leq 2)$ at 300 K and 550 K.*

## 3.3 Quaternary System

### *3.3.1 Structural analysis*

X-ray diffraction is the greatest and significant method for the investigation of crystal structure of nanomaterials. With the purpose, to check the purities of different phases of compound peaks in XRD figures, as per revealed in **Figure 9**. It is authenticated that the XRD design of all these samples are fine unchanging with the literature and has been recognized that the crystal structure scheme is isostructural with reference data of $Tl_9GdTe_6$ and $Tl_9BiTe_6$ having tetragonal crystal structure with the space group symbol of 4/cm. The SEM shows, morphological structure at the 100 nm scale. The energy dispersion X-ray diffractometer show the concentrate composition of the compound in **Figure 10**.

## 3.4 Physical properties

### *3.4.1 Electrical conductivity measurements*

The temperature variations of electrical conductivity of quaternary compounds are revealed in **Figure 11**. The conductivity experiential for the entire samples are studied here, decreases with increasing temperature, representing the degenerate semiconductor performance because of positive temperature coefficient, subsequent from the phonons scattering of charge carriers and grains boundaries effects.

| Sample | Power factor ($\mu Wcm^{-2}K^{-2}$)at 300 K | Power factor ($\mu Wcm^{-2}K^{-2}$) at 550 K |
|---|---|---|
| $Tl_9Sn_1Te_6$ | 1.9 | 4.505 |
| $Tl_{8.75}Sn_{1.25}Te_6$ | 2.637 | 9.234 |
| $Tl_{8.50}Sn_{1.50}Te_6$ | 6.755 | 9.579 |
| $Tl_{8.25}Sn_{1.75}Te_6$ | 7.574 | 10.135 |
| $Tl_8Sn_2Te_6$ | 3.638 | 9.777 |

**Table 4.**
*Power factor of* $Tl_{10-x}Sn_xTe_6(1 \leq x \leq 2)$ *at* 300 K *and* 550 K.

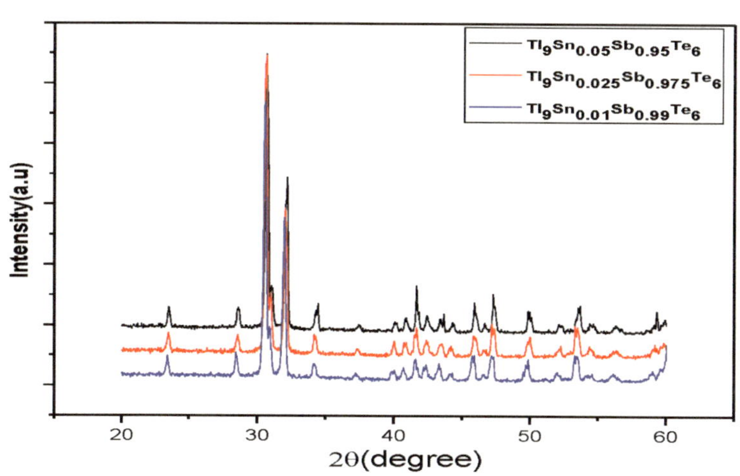


**Figure 9.**
*XRD data of* $Tl_9(SnSb)_1Te_6$ *with Sn = 0.01, 0.025 and 0.05.*

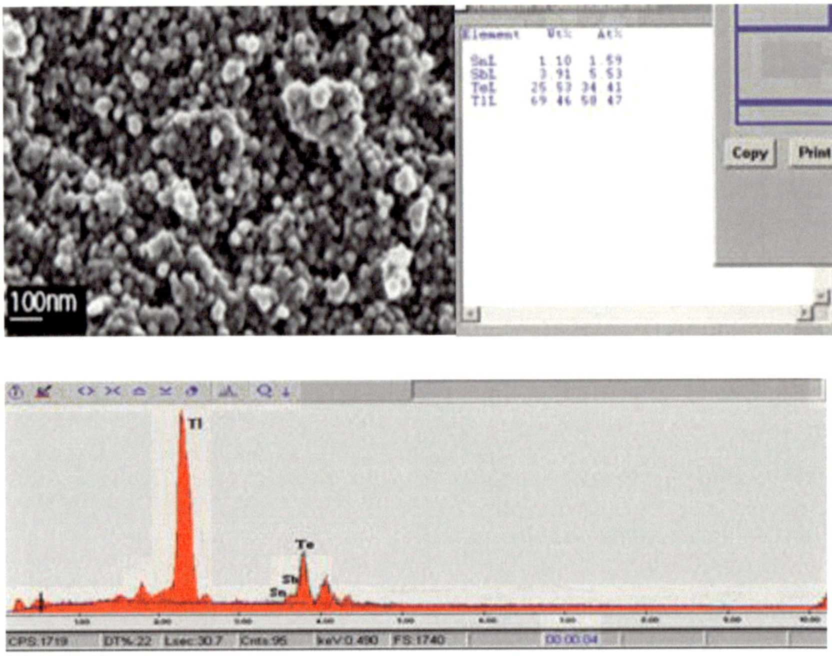


**Figure 10.**
*SEM and EDX image of $Tl_9(SnSb)_1Te_6$.*

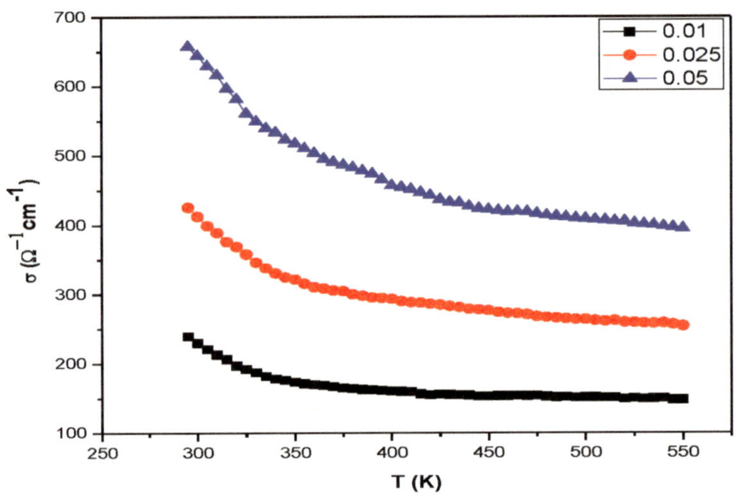


**Figure 11.**
*Electrical conductivity measurements at different high temperature.*

An increasing "*x*" value (i.e. increasing the Sn deficiency) is predictable to increase the number of holes, which is experimental detected. The smaller temperature need may be produced by (less temperature dependence) more grain boundary scattering. No systematic trend was found in the variation of the electrical resistivity for samples $Tl_9Sb_{1-x}Sn_xTe_6$ (x = 0.01, 0.025, and 0.05) with "Sn" concentration. The low electrical conductivity in the pressure less sintered sample may be caused by means of the oxide impurity phase in the grain boundary and the number of the grain boundary. The *Sn* doping level and grain boundary resistance may play significant part for increasing electrical conductivity.

#### *3.4.2 Seebeck coefficient (S) analyses*

To examine the influence of decrease of the charge carriers in thermal and transport features, Sn content was increased in $Tl_9Sb_{1-x}Sn_xTe_6$ (x = 0.01, 0.025, and 0.05) by means of replacing Sb atoms conferring to the formula. The temperature variation as a function of the Seebeck coefficient (S) for the $Tl_9Sb_{1-x}Sn_xTe_6$ (x = 0.01, 0.025 and 0.05) compounds are revealed in **Figure 12**. The Seebeck coefficient was measured in the temperature gradient of 1 K. The positive Seebeck coefficient increases easily with increasing temperature from 300 K to 550 K, for all compounds in mainly for p-type semiconductors having high charge carrier concentration. It is understandable that all the samples display positive Seebeck coefficient for the whole temperature range, signifying that the p-type (hole) carrier's conduction controls the thermoelectric transportation in these compounds. When the amount of Sn increased from 0.01 to 0.05, the Sn doping is supposed to increase the carrier's density. Though, the smaller grains upon Sn doping are thought to be talented to improve the electron scattering, yielding an increase of the Seebeck coefficient and effective mass.

#### *3.4.3 Power factor calculation and analysis*

To improve the power factor (PF = $S^2\sigma$) for these compounds, we require to decouple the electrical conductivity from the Seebeck coefficient, typically inversely proportional to each other in these systems. The key contribution in the "PF" originates from the Seebeck coefficient, so we must design the materials such that their "*S*" should be improved. The power factors calculated from the electrical conductivity "σ" and the square of Seebeck coefficient "S," gotten for $Tl_9Sb_{1-x}Sn_xTe_6$ compounds with x = 0.01, 0.025 and 0.05 are showed in **Figure 13**. The power factor increases with increasing temperature for all these compounds. The doping concentration demonstrations a systematic effect on the power factor as increasing the doping concentration, the power factor is increases. The

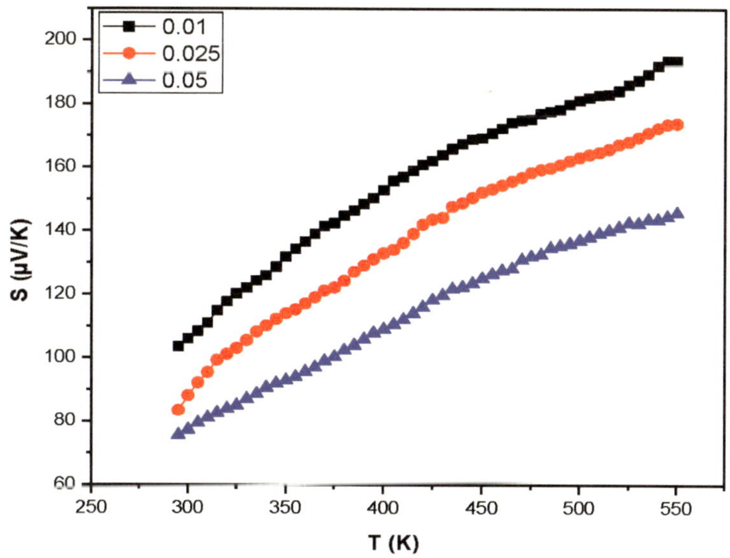


**Figure 12.**
*Seebeck coefficient measurements at different concentration and high temperature (0.01, 0.025, 0.05).*

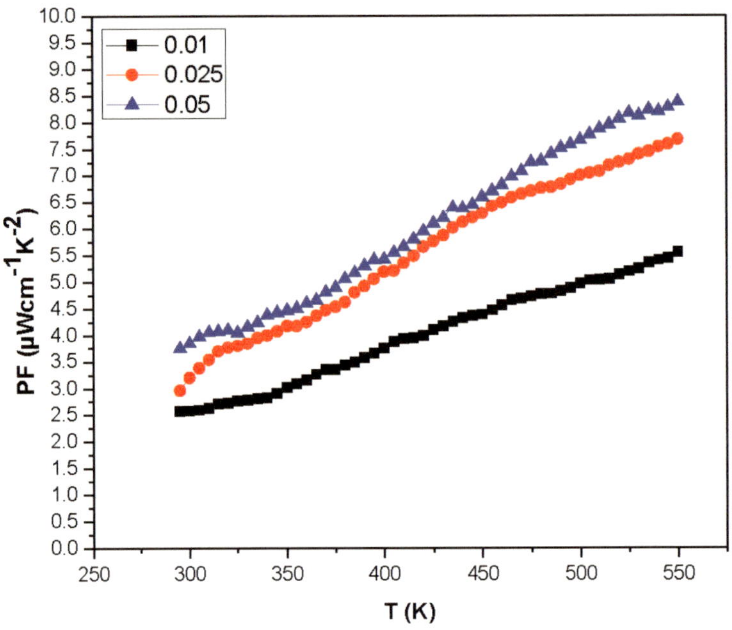


**Figure 13.**
*The power factor* (PF) *of* $Tl_9Sb_{1-x}Sn_xTe_6$ *with* $x = 0.01, 0.025, 0.05$.

$Tl_9Sb_{0.95}Sn_{0.05}Te_6$ compound showed the highest value 8.37 (μWtt-cm$^{-1}$-K$^{-2}$) of "PF" at 550 K and 3.75 (μWtt-cm$^{-1}$-K$^{-2}$) at 295 K. The lowest "PF" were experiential for $Tl_9Sb_{0.99}Sn_{0.01}Te_6$ compound which have values of 5.55 (μWtt-cm$^{-1}$-K$^{-2}$) at 550 K and 2.56 (μWtt-cm$^{-1}$-K$^{-2}$) at 295 K. As deliberated earlier, an increasing the "Sn" contents are probable to increase the number of holes and the dominant charge carriers.

## 4. Conclusion

In this study, the ternary and quaternary Tellurium Telluride chalcogenides, $Tl_{10\text{-}x\text{-}y}A_xB_yTe_6$ nanoparticles, with different types of dopants (A = Sb, and B = Sn) and with different concentration of Sn has been introduced to synthesize new materials by co-precipitation techniques and explored their structural, electrical and thermal properties has been analyzed in details. The structural investigation revealed that $Tl_{10\text{-}x}Sn_xTe_6$ is isostructural with $Tl_5Te_3$ with a same space group I4/mcm. All peaks are corresponding to their respective element, and no extra peaks are observed, which shows that we got a correct crystal structure for our design materials and also shows that no impurities or dislocation in the sample has been observed. An energy dispersive X-ray spectroscopy was used for the confirmation of elemental and compositional ratio of all the samples studied here. The electrical characterizations shows that parent compounds behaves like a semiconductor, but increasing the Sn contents, this materials tend toward the metallic properties, which show that increasing the temperature the electrical conductivity will decreases. The electrical characterizations show that parent compounds behaves like a semiconductor, but increasing the *Sn* contents, this nanomaterials tend to metallic phase, which display that increasing the temperature the electrical conductivity will decreases at higher temperature.

All samples exhibited positive S values, increasing Sn-filling, the Seebeck coefficient increased due to increase in its metallic behavior and low thermal conductivity. By increasing the temperature, the Seebeck coefficient was increased and the highest Seebeck coefficient was observed for $Tl_8Sn_2Te_6$; $S = +157\mu VK^{-1}$. Consequently, power factor was enhanced and increased with high Sn concentration up to $Sn = 1.75$ and the maximum power factor $(PF = 7.579\mu Wcm^{-2}K^{-2})$ was observed for $Tl_{8.25}Sn_{1.75}Te_6$. The reduction in the power factor for $Tl_8Sn_2Te_6$ is due to their low electrical conductivity. The thermopower is positive in the whole temperature range studied here, which is increasing with increase in temperature, representing that the nanoparticles under study is hole conduction dominated. For higher concentrations of *Sn*, the Seebeck coefficient of the doped tellurium telluride is decreasing because of increasing the holes concentration which in turns increasing the electron scattering in this doped chalcogenide system. However, the smaller grains upon Sn concentrations will improve the electron scattering, resulting increase in thermos-power. Therefore, power factor was improved and increased with high "Sn" concentration up to Sn = 0.05 and the maximum power factor ($PF = 8.37\ \mu Wcm^{-1}K^{-2}$) was observed for $Tl_9Sb_{0.95}Sn_{0.05}Te_6$. This enhanced power factor will improve the thermoelectric efficiency and results decent thermoelectric applications, which is the key goal of this study. At the end, we are going to accomplish that this work is the finest example of enhancing dopants concentration to attain required thermoelectric properties in "*Sn*" doped $Tl_9Sb_{1\text{-}x}Sn_xTe_6$ chalcogenide system. To understand our results, we start by final the basic understanding of the metallic long-range interactions due to do the injections of charge carriers concentration, and semiconducting frustration effects fore-most to metallic like conduct in these chalcogenides.

## Author details

Wiqar Hussain Shah* and Waqas Muhammad Khan
Department of Physics, Faculty of Basic and Applied Sciences, International Islamic University, Islamabad, Pakistan

*Address all correspondence to: wiqar.hussain@iiu.edu.pk

## References


[1] Kurosaki K, Uneda H, Muta H, Yamanaka S. Thermoelectric properties of thallium antimony telluride. Journal of Alloys and Compounds. 2004:43-48

[2] Assoud A, Soheilnia N, Kleinke H. Crystal structure, electronic structure and physical properties of the new low-valent thallium silicon telluride Tl6Si2Te6 in comparison to Tl6Ge2Te6. Journal of Solid State Chemistry. 2006; **179**:2707-2713

[3] Kurosaki K, Goto K, Muta H, Yamanaka S. Fabrication and thermoelectric properties of Ag9TlTeX (X=5:0~6:0). Materials Transactions. 2007;**48**(8):2083-2087

[4] Heremans JP, Jovovic V, Toberer ES, Saramat A, Kurosaki K, Charoenphakdee A, et al. Enhancement of thermoelectric efficiency in PbTe by distortion of the electronic density of states. Science. 2008;**321**:554-557

[5] Sankar CR, Sanasy SB, Assoud A, Kleinke H. Syntheses, crystal structures and thermoelectric properties of two new thallium tellurides: Tl4ZrTe4 and Tl4HfTe4. Journal of Materials Chemistry. 2010;**20**:7485-7490

[6] Sankar CR, Guch M, Assoud A, Kleinke H. Structural, thermal, and physical properties of the thallium zirconium telluride Tl2ZrTe3. Chemistry of Materials. 2011;**23**: 3886-3891

[7] Raj C, Savitree BS, Holger K. Thermoelectric properties of TlGdQ2(Q = Se, Te) and Tl9GdTe6. Journal of Electronic Materials. 2012;**41**(6): 1662-1666

[8] Sanasy SB. Thermoelectric properties of Tl10-xLnxTe6, with Ln=Ce, Pr, Nd, Sm, Gd, Tb, Dy, and Er, and 0.25<x< 1.32. Journal of Alloys and Compounds. 2013:126-134

[9] Kim KT, Ha GH. Fabrication and enhanced thermoelectric properties of alumina nanoparticle-dispersed Bi0.5Sb1.5Te3 matrix composites. Journal of Nanomaterials. 2013;**2013**:1-6

[10] Zhang Q, Liao B, Lan Y, Lukas K, Liu W, Esfarjani K, et al. High thermoelectric performance by resonant dopant indium in nanostructured SnTe. Applied Physical Sciences. 2013;**110**(33): 13261-13266

[11] Bali A, Wang H, Snyder GJ, Mallik RC. Thermoelectric properties of indium doped PbTe1-ySey alloys. Journal of Applied Physics. 2014;**116**: 033707

[12] Kuropatwa BA, Guo Q, Assoud A, Kleinke H. Optimization of the telluride Tl10–x–ySnxBiyTe6 for the thermoelectric energy conversion. Journal of Inorganic and General Chemistry. 2014;**640**:774-780

[13] Dresselhaus MS, Chen G, Tang MY, Yang RG, Lee H, Wang DZ, et al. New directions for low-dimensional thermoelectric materials. Advanced Materials. 2007;**19**(8):1043-1053

Chapter 5

# Electrical Conductivity of Molten Salts and Ionic Conduction in Electrolyte Solutions

*Shigeru Tamaki, Shigeki Matsunaga and Masanobu Kusakabe*

## Abstract

A microscopic description for the partial DC conductivities in molten salts has been discussed by using a Langevin equation for the constituent ions. The memory function $\gamma(t)$ can be written as in the form of a decaying function with time. In order to solve the mutual relation between the combined-velocity correlation functions $Z_{\sigma}^{\pm}(t)$ and the memory function $\gamma(t)$ in a short time region, a new recursion method is proposed. Practical application is carried out for molten NaCl by using MD simulation. The fitted function is described by three kinds of Gaussian functions and their physical backgrounds are discussed. Also the electrical conductivity in aqueous solution of electrolyte has been obtained, based on a generalized Langevin equation for cation and anion in it. This treatment can connect and compare with the work of computer simulation. The obtained results for concentration dependence of electrical conductivity are given by a function of the square root of concentration. The electrophoretic effect and the relaxation one are also discussed.

**Keywords:** conductivity of molten salts, conductivity of electrolytic solution, Langevin equation, MD simulation

## 1. Introduction

The phenomena of transport properties in ionic liquids are of great important in the industrial science and technology, as well as in physics and chemistry. In connection with these, a number of experimental and theoretical studies have been published until the present time [1–3]. Ionic liquids are mainly classified into two categories; one is a group of molten salts and the other is a large number of electrolytic solutions, in particular, aqueous solutions of electrolytes.

In the case of molten salts, Sundheim discovered that the ratio of the partial conductivities of cation and anion were always equal to their inverse mass ratio, namely, $\sigma^{+}(DC)/\sigma^{-}(DC) = m^{-}/m^{+}$ [4].

Later on, this golden rule or a unified rule was theoretically explained by our group [5–9]. Detailed procedure will be shown in what follows.

Paralleling to above discovery, a number of scientific studies in molten salts have been developed from 1960s by several researchers [10, 11].

In order to study the structural and transport properties in molten salts, experimental investigations and molecular dynamics simulations have also been carried out from mid-70s of the last century [12–16].

Following to these, we have been engaged in the study of transport properties in molten salts [6–9, 17]. We have carried out a theoretical study on the electrical conductivity of molten salts, starting from the Langevin equation and the velocity correlation functions for the constituent ions. Subsequently this treatment was successful to obtain the golden rule $\sigma^{+}/\sigma^{-} = m^{-}/m^{+}$ in a microscopic view point.

It remains, however, unclear how the adopted Langevin equation can be effectively solved within a short time region, under an appropriate memory function, because our former theory was only successful to get the partial conductivities.

We like to discuss more generally the correlation between the velocity correlation functions incorporated with the partial DC conductivities and some of useful memory functions which are closely related to the friction constants acting on cations and anions in molten salts.

Preceding the investigation for molten salts, on the other hand, there have been a number of studies for ionic solutions since the discovery of Faraday, in which a typical example is electrolytic solution. During such long-termed history of electrochemistry, it was well established by Kohlrausch that the experimental results on the ionic conductivities in dilute electrolytic solutions indicated the law of independent migration of ions, $\Lambda_c = \Lambda_0 - kc^{1/2}$, where $\Lambda_0$ being the conductivity in the dilute limit and c the concentration and k the constant specified by the electrolyte dissolved in water.

The beginning of the modern aspect, in particular, on the thermodynamic and transport properties in electrolytic solutions might be originated from Debye-Hückel theory [18].

In order to explain the ionic conductivity in electrolytic solution, successful works following to Debye-Hückel theory have been reported by Onsager [19], Prigogine [20], and Fuoss and his co-worker [21]. In these theories, $\Lambda_0$ is treated by the Stokes law and the concentration dependence is mainly explained by the electrophoretic effect and relaxation one. Therefore, these treatments are based on a kind of mixing of the microscopic and partially macroscopic view point.

Starting from the Liouville equation, statistical mechanics of irreversible process for the ionic conductivity in electrolytic solution have been developed by Davis and Résibois [22] and Friedman [23], although they did not derive any explicit expressions for the friction constant in terms of inter-particle interactions.

It has been required to investigate the static and dynamic properties of dissolved ions in aqueous solutions from the microscopic view point. Along this requirement, the technique of molecular dynamic simulation has been applied, using some qualified inter-particle potentials. Various theoretical attempts have been recently tried to establish the dynamical behaviors of dissolved ions in these solutions, which is able to discuss parallel with results obtained by MD simulation [24–26].

Chandra and Bagchi [27] have developed a new theoretical approach to study the ionic conduction in electrolytic solutions, based on the combination of the mode coupling theory and the generalized Langevin equation, and they were successful to obtain the Onsager equation. However, there still remains the task to obtain how to derive the theoretical formula for $\Lambda_0$ in terms of inter-particle potentials and corresponding pair distribution functions.

We will apply the linear response theory for the electrolytic solution and to obtain $\Lambda_0$ and the concentration dependence of the conductivity in terms of pair-wise potentials and pair distribution functions among ions and water molecules, which can compare parallel with dynamical properties of MD simulation [28].

In addition, we will also clarify how the electrophoretic and relaxation effects treated by many researchers are explained in a microscopic view point.

From these, we will see what is similar and what is different for the case of molten salts and that of electrolytic solutions.

## 2. Generalized Langevin equations for the cation and anion in a molten salt

Let us consider a molten salt composed of the density $n^+ = n^- = n_0$ $(= N/V_0)$, of the constituent ion's masses $m^+$ and $m^-$, and of the charge $z^+ = -z^- = z = 1$, where N being the total number of cation and/or anion in the volume $V_0$.

A golden rule, $\sigma^+(DC)/\sigma^-(DC) = m^-/m^+$, can be obtainable from a generalized Drude theory, as a law of motion under an electric field [5].

As an extension, the generalized Langevin equation for an arbitrary cation or anion in the system under an external field **E** is written as follows:

$$m^{\pm}d\mathbf{v}_i^{\pm}(t)/dt = -m^{\pm}\int_{-\infty}^{t}\xi^{\pm}(t-t')\mathbf{v}_i^{\pm}(t')dt' + \mathbf{R}_i^{\pm}(t) + z^{\pm}e\mathbf{E} \quad (1)$$

where $\xi^{\pm}(t)$ and $\mathbf{R}_i^{\pm}(t)$ are the retarded friction function in relation to the friction force and the random fluctuating force, acting on the cation or anion *i*, respectively.

After taking the ensemble average, equations of time evolution based on Eq. (1) in respect to the partial ionic conductivities are then written as follows:

$$m^{\pm}d<\mathbf{v}_i^{\pm}(t)\mathbf{v}_j^{\pm}(0)>/dt = -m^{\pm}\int_{-\infty}^{t}<\xi^{\pm}(t-t')\mathbf{v}_i^{\pm}(t')\mathbf{v}_j^{\pm}(0)>dt' \quad (\text{for } i=j \text{ and } i\neq j) \quad (2)$$

and

$$m^{\pm}d<\mathbf{v}_i^{\pm}(t)\,\mathbf{v}_k^{\mp}(0)>/dt = -m^{\pm}\int_{-\infty}^{t}<\xi^{\pm}(t-t')\mathbf{v}_i^{\pm}(t')\mathbf{v}_k^{\mp}(0)> \quad (\text{for } i\neq k) \quad (3)$$

And the equation of time evolution in relation to the diffusion constants of constituent ions is written as follows:

$$m^{\pm}d<\mathbf{v}_i^{\pm}(t)\mathbf{v}_i^{\pm}(0)>/dt = -m^{\pm}\int_{-\infty}^{t}<\xi^{\pm}(t-t')\mathbf{v}_i^{\pm}(t')\mathbf{v}_i^{\pm}(0)>dt' \quad (4)$$

As was previously illustrated [9], the retarded friction function $\xi^{\pm}(t)$ cannot be independent for the averaging procedure and we have to define new memory functions as follows:

$$<\xi^{\pm}(t-t')\mathbf{v}_i^{\pm}(t')\mathbf{v}_j^{\pm}(0)> = \gamma_\sigma^{\pm}(t)<\mathbf{v}_i^{\pm}(t')\mathbf{v}_j^{\pm}(0)> \quad (\text{for } i=j \text{ and } i\neq j) \quad (5)$$

and

$$<\xi^{\pm}(t-t')\mathbf{v}_i^{\pm}(t')\mathbf{v}_k^{\mp}(0)> = \gamma_\sigma^{\pm}(t) <\mathbf{v}_i^{\pm}(t')\mathbf{v}_k^{\mp}(0)> \quad (\text{for i} \neq \text{k}) \tag{6}$$

While, in the case of diffusion constants of constituent ions, that is, $\mathbf{E} = 0$, we can define

$$<\xi^{\pm}(t-t')\mathbf{v}_i^{\pm}(t')\mathbf{v}_i^{\pm}(0)> = \gamma_D^{\pm}(t) <\mathbf{v}_i^{\pm}(t')\mathbf{v}_i^{\pm}(0)> \tag{7}$$

It is emphasized that the memory functions $\gamma_\sigma^{\pm}(t)$ is not equal to $\gamma_D^{\pm}(t)$ as shown in previous paper [9]. In other words, the retarded friction function, $\xi^{\pm}(t-t')$, is a kind of vector function and is varied with the environment such as the existence of electric field **E**. Therefore, the memory function is varied in accordance with what sort of evolution is considered in the time-dependent correlation function [29].

Assuming that the ensemble average for the fluctuating force is zero and if we apply the following electric field,

$$\mathbf{E}(t) = \text{Re } \mathbf{E}_0 \exp(\text{i}\omega t) \tag{8}$$

where Re means the real part and ω is the angular frequency, then the averaged ion's velocity induced by this external filed is equal to

$$<\mathbf{v}_i^{\pm}(t)> = \text{Re } \mu^{\pm}(\omega) z^{\pm} e\mathbf{E}(t) \tag{9}$$

where $\mu^{\pm}(\omega)$ is the mobility of cation or anion.

Putting (9) into the equation of motion (1) after taking the ensemble average, we have

$$\mu^{\pm}(\omega) = (1/m^{\pm})\left[1/\{\text{i}\omega + \tilde{\gamma}^{\pm}(\omega)\}\right] \tag{10}$$

where

$$\tilde{\gamma}^{\pm}(\omega) = \int_0^{\infty} \gamma^{\pm}(t) \exp(-\text{i}\omega t)\, \text{d}t \tag{11}$$

Therefore, the current density is written as follows:

$$\text{j}^{\pm}(t) = \text{n}z^{\pm 2}e^2 <\mathbf{v}_i^{\pm}(t)> = \text{Re } \text{n}z^{\pm 2}e^2\mu^{\pm}(\omega)\mathbf{E}(t) \tag{12}$$

The partial conductivity is, then, equal to

$$\sigma^{\pm}(\omega) = \text{n}z^{\pm 2}e^2\mu^{\pm}(\omega) = \left(\text{n}z^{\pm 2}e^2/\text{m}^{\pm}\right)\left\{1/\left(\text{i}\omega + \tilde{\gamma}^{\pm}(\omega)\right)\right\} \tag{13}$$

and in the limit of ω = 0,

$$\sigma^{\pm}(\text{DC}) = \text{n}z^{\pm 2}e^2\mu^{\pm}(0) = \left\{\text{n}z^{\pm 2}e^2/\text{m}^{\pm}\tilde{\gamma}^{\pm}(0)\right\} \tag{14}$$

Therefore, $\tilde{\gamma}^{\pm}(0)$ is equal to the effective friction constant acting on each ion. According to our previous studies [7–9], the following relation was recognized:

$$\tilde{\gamma}^{+}(0) = \tilde{\gamma}^{-}(0) \equiv \tilde{\gamma}(0) \tag{15}$$

where $\tilde{\gamma}(0)$ is expressed as follows:

$$\tilde{\gamma}(0) = \left(\alpha^0/3\mu\right)^{1/2}, \quad (1/\mu) = (1/m^+) + (1/m^-) \tag{16}$$

and

$$\alpha^0 = n\int_0^\infty \left[\partial^2\phi^{+-}(r)/\partial r^2 + (2/r)\{\partial\phi^{+-}(r)/\partial r\}\right] g^{+-}(r)\cdot 4\pi r^2 dr \tag{17}$$

$\phi^{+-}(r)$ and $g^{+-}(r)$ in this equation are the inter-ionic potential between cation and anion and the corresponding pair distribution function, respectively.

Therefore, we have a golden rule for the partial conductivities in a microscopic scale as follows:

$$\sigma^+(DC)/\sigma^-(DC) = m^-/m^+ \tag{18}$$

In the following sections, as a numerical example, the MD simulation on molten NaCl at 1100 K is often utilized, for which the interionic potential functions suggested by Tosi and Fumi [30] for a study of solid alkali halides are applied. In order to make sure that the Tosi-Fumi potential for NaCl can be valid in the liquid state, we have estimated the partial pair distribution functions of molten NaCl liquid, $g_{ij}(r)$ (i,j = $Na^+$, $Cl^-$) as shown in **Figure 1**, which agree with those of experimental results obtained by Edwards et al. [31].

Using these $g_{ij}(r)$, we have also estimated the total neighboring numbers around arbitrary ions located at the distance r, which describe as $n_{ij} = 4\pi\int_0^r r^2 dr$, as shown in **Figure 2a–c**.

The nearest neighbor number is defined as $n_{ij}(r_1)$, where $r_1$ is the position of the first minimum of $g_{ij}(r)$.

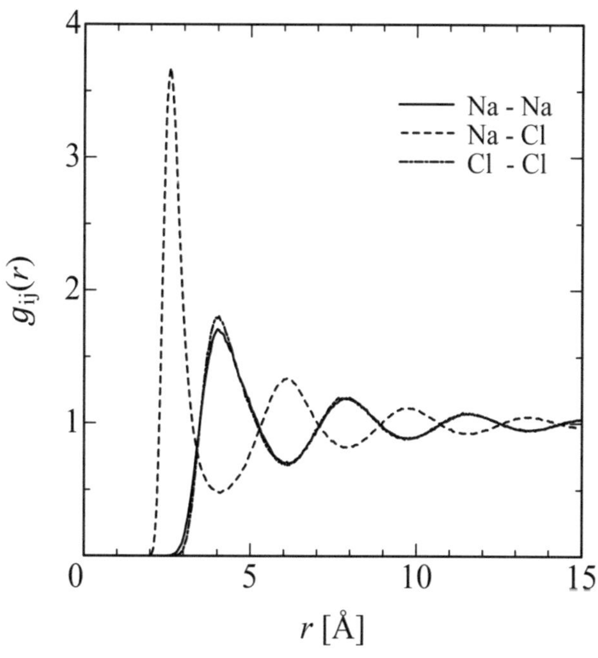


**Figure 1.**
*Pair distribution functions, $g_{ij}(r)$, for molten NaCl at 1148 K, obtained by MD simulation.*

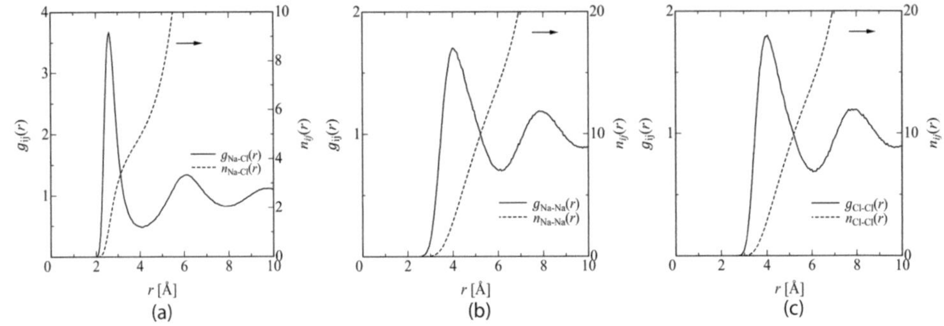


**Figure 2.**
*(a)* gNa-Cl(r) *and* nNa-Cl(r) *for molten NaCl at 1148 K, obtained by MD simulation. (b)* gNa-Na(r) *and* nNa-Na(r) *for molten NaCl at 1148 K, obtained by MD simulation. (c)* gCl-Cl(r) *and* nCl-Cl(r) *for molten NaCl at 1148 K, obtained by MD simulation.*

Then, the nearest neighbors around a $Na^+$ are nearly equal to 5.0, since the distance $r_1$ is taken at the minimum position of $g_{Na\text{-}Cl}(r)$ as shown in **Figure 2a**.

The application of Tosi-Fumi potentials in the MD simulations for viscosity and electrical conductivity is also valid to reproduce their experimental results [5].

Therefore, the following MD simulations for molten NaCl must be reliable to see their microscopic view.

## 3. Linear response theory for the partial conductivities

On the other hand, according to our previous investigations [6–9, 17, 29], the partial DC conductivities $\sigma^+(DC)$ and $\sigma^+(DC)$ are expressed as follows,

$$\sigma^+(DC) = \sigma^{++} + \sigma^{+-} = (1/3k_BT)\int_0^\infty <\mathbf{j}^+(t)\,\mathbf{j}(0)> dt \tag{19}$$

$$\sigma^-(DC) = \sigma^{--} + \sigma^{+-} = (1/3k_BT)\int_0^\infty <\mathbf{j}^-(t)\,\mathbf{j}(0)> dt \tag{20}$$

where

$$\sigma^{\pm\pm} = (1/3k_BT)\int_0^\infty <\mathbf{j}^\pm(t)\,\mathbf{j}^\pm(0)> dt \tag{21}$$

$$\sigma^{+-} = (1/3k_BT)\int_0^\infty <\mathbf{j}^+(t)\,\mathbf{j}^-(0)> dt \tag{22}$$

and

$$\mathbf{j}(t) = \mathbf{j}^+(t) + \mathbf{j}^-(t) \tag{23}$$

where

$$\mathbf{j}^+(t) = \sum_{i=1}^{n} z^+ e\mathbf{v}_i{}^+(t), \quad \mathbf{j}^-(t) = \sum_{k=1}^{n} z^- e\mathbf{v}_k{}^-(t) \tag{24}$$

Considering the ensemble averages of (19) and (20), it is convenient to define the velocity correlation functions $Z_\sigma{}^+(t)$ and $Z_\sigma{}^-(t)$ as follows:

$$Z_\sigma^{\ +}(t) \equiv <\mathbf{v}_i^{\ +}(t)\,\mathbf{v}_j^{\ +}(0)> - <\mathbf{v}_i^{\ +}(t)\,\mathbf{v}_k^{\ -}(0)> \tag{25}$$

and

$$Z_\sigma^{\ -}(t) \equiv <\mathbf{v}_k^{\ -}(t)\,\mathbf{v}_l^{\ -}(0)> - <\mathbf{v}_i^{\ +}(t)\,\mathbf{v}_k^{\ -}(0)> \tag{26}$$

where < > means the ensemble average.

Using (25) and (26), the partial DC conductivities (19) and (20) are written, respectively, as follows:

$$\sigma^+(\mathrm{DC}) = \left(\mathrm{nz^2e^2/3k_BT}\right)\int_0^\infty Z_\sigma^{\ +}(t)\mathrm{d}t \tag{27}$$

$$\sigma^-(\mathrm{DC}) = \left(\mathrm{nz^2e^2/3k_BT}\right)\int_0^\infty Z_\sigma^{\ -}(t)\mathrm{d}t \tag{28}$$

On the other hand, combining Eqs. (25) or (26) and (1), we have

$$\partial\{Z_\sigma^{\ +}(t)\}/\partial t = -\int_0^t \gamma^+(t-s)Z_\sigma^{\ +}(s)\mathrm{d}s \tag{29}$$

and/or

$$\partial\{Z_\sigma^{\ -}(t)\}/\partial t = -\int_0^t \gamma^-(t-s)Z_\sigma^{\ -}(s)\mathrm{d}s \tag{30}$$

Taking the Laplace transformation of $\partial\{Z_\sigma^{\ +}(t)\}/\partial t$ in (29) as follows,

$$\mathcal{L}\left[\partial\{Z_\sigma^{\ +}(t)\}/\partial t\right] \equiv \int_0^\infty \exp(-\mathrm{i}\omega t)[\partial\{Z_\sigma^{\ +}(t)/\partial t\}]\mathrm{d}t \tag{31}$$

$$= \left[\exp(-\mathrm{i}\omega t)Z_\sigma^{\ +}(t)\right]_0^{\ \infty} + \mathrm{i}\omega\int_0^\infty \exp(-\mathrm{i}\omega t)Z_\sigma^{\ +}(t)\mathrm{d}t \tag{32}$$

$$= -Z_\sigma^{\ +}(0) + \mathrm{i}\omega\tilde{Z}_\sigma^{\ +}(\omega) \tag{33}$$

Here, we have used an evident condition $Z_\sigma^{\ +}(t = \infty) = 0$.

On the other hand, the right hand side of (29) is given by the following expressions:

$$\mathcal{L}\left\{-\int_0^\infty \gamma^+(s)Z_\sigma^{\ +}(t-s)\mathrm{d}s\right\} = -\int_0^\infty \exp\{-\mathrm{i}\omega(t-s)\}\gamma^+(t-s)\,\mathrm{d}(t-s)$$
$$\int_0^\infty \exp(-\mathrm{i}\omega s)Z_\sigma^{\ +}(s)\mathrm{d}s = -\tilde{\gamma}^+(\omega)\,\tilde{Z}_\sigma^{\ +}(\omega) \tag{34}$$

Therefore we have,

$$-Z_\sigma^{\ +}(0) + \mathrm{i}\omega\tilde{Z}_\sigma^{\ +}(\omega) = -\tilde{\gamma}^+(\omega)\,\tilde{Z}_\sigma^{\ +}(\omega) \tag{35}$$

$$\tilde{Z}_\sigma^{\ +}(\omega) = Z_\sigma^{\ +}(0)/\{\mathrm{i}\omega + \tilde{\gamma}^+(\omega)\} = (3\mathrm{k_BT/m^+})/\{\mathrm{i}\omega + \tilde{\gamma}^+(\omega)\} \tag{36}$$

In a similar way, we have,

$$\tilde{Z}_\sigma^{-}(\omega) = Z_\sigma^{\ -}(0)/\{\mathrm{i}\omega + \tilde{\gamma}^-(\omega)\} = (3\mathrm{k_BT/m^-})/\{\mathrm{i}\omega + \tilde{\gamma}^-(\omega)\} \tag{37}$$

If an appropriate memory function $\gamma(t)$, which is valid for both cation and anion in the system, is considered and its Laplace transformation is inserted into either (36) or (37), then we can get the partial AC conductivities.

## 4. Microscopic representation for the $Z_\sigma^+(t)$ and $Z_\sigma^-(t)$ in a molten salt

We have already shown the microscopic expressions for $Z_\sigma^+(t)$ and $Z_\sigma^-(t)$ as Taylor expansion forms in a molten salt in which the inter-ionic potential between cation and anion and the corresponding pair distribution function are concerned by Koishi et al. [7]. In these combined velocity correlation functions, it can be shown that the odd power terms of the time $t$ have vanishing coefficients which, it turns out, is related to the fact that any positions and their differentiations with time are uncorrelated in an ensemble average. In facts, the velocity auto-correlation function can be expressed in terms of even powers of the time $t$ [32, 33].

The short-time expansion forms of $Z_\sigma^+(t)$ and $Z_\sigma^-(t)$ are actually shown in the following forms:

$$Z_\sigma^+(t) = (3k_BT/m^+)\left[1 - (t^2/2)(\alpha^0/3\mu) + (\mathrm{over}\,t^4)\right] \tag{38}$$

and

$$Z_\sigma^-(t) = (3k_BT/m^-)\left[1 - (t^2/2)(\alpha^0/3\mu) + (\mathrm{over}\,t^4)\right] \tag{39}$$

Thus, the partial conductivities for cation and anion in a molten salt are written as in the following Kubo-formulae:

$$\sigma^+(DC) = (n_0e^2/m^+)\int_0^\infty \{1 - (t^2/2)(\alpha^0/3\mu) + (\mathrm{over}\,t^4)\}dt \tag{40}$$

and

$$\sigma^-(DC) = (n_0e^2/m^-)\int_0^\infty \{1 - (t^2/2)(\alpha^0/3\mu) + (\mathrm{over}\,t^4)\}dt \tag{41}$$

Using (14), (16), (40) and (41), we have a very interesting relation written in the following form:

$$1/\tilde{\gamma}(0) = \int_0^\infty \left\{1 - (t^2/2)(\tilde{\gamma}(0))^2 + (\mathrm{over}\,t^4)\right\}dt \tag{42}$$

However, it is generally difficult to obtain $Z_\sigma^\pm(t)$ from appropriate memory functions, by using the well-known method in statistical mechanics [33].

Under these circumstances, we explore a new method to solve Langevin Eqs. (29) and (30), in order to clarify a detailed correlation between $\gamma(t)$ and $Z_\sigma^\pm(t)$ within the short time region, which will be shown in later section.

## 5. Method of continued-fraction based on Mori formulae

Many years ago, Mori [34, 35] had generalized the Langevin equation starting from the Hamilton's canonical equation of motion in a system of a monatomic liquid with the component's mass as $m$. Along his theory, Copley and Lovesey [36] have concluded that the memory function in the generalized Langevin equation could be expressed as follows:

$$\partial\{\gamma_n(t)\}/\partial t = -\int_0^t \gamma_{n+1}(t-s)\gamma_n(s)ds \quad n = 1, 2, 3, \ldots \tag{43}$$

where $\gamma_n(t)$ is the n-th stage memory function and the first stage memory function is equal to $\gamma(t)$ in Eqs. (29) and (30). The Fourier-Laplace transform of the above equation provides the following continued-fraction representation,

$$\tilde{\gamma}_n(\omega) = -\delta_n/[\omega + \tilde{\gamma}_{n+1}(\omega)] \tag{44}$$

where the Mori coefficient $\delta_n$ is equal to $\gamma_n(0)$.

The method of Copley and Lovesey [36] was able to express the short time expansion for the velocity correlation function $Z(t)$ ($= < \mathbf{v}_i(t)\ \mathbf{v}_j(0)>$) described as in the following form:

$$Z(t) = Z_0\{1 - (t^2/2!)Z_2 + (t^4/4!)Z_4 - (t^6/6!)Z_6 + \ldots\} \tag{45}$$

Thus, they provided the following relations if several $\delta_n$'s are known:

$$Z_0 = (3k_BT/m), \quad Z_2 = Z_0\delta_1, \quad Z_4 = Z_0\delta_1(\delta_1 + \delta_2), \quad Z_6 = Z_0\{(\delta_1 + \delta_2) + \delta_2\delta_3\}, \ldots \tag{46}$$

Therefore, the problem is ascribed to the derivation of $\delta_n$'s. Because of a hard task in such repeating calculations, it is difficult to obtain a number of $\delta_n$'s. However, several applications along these procedures have been carried out [37, 38].

Instead of the method of continued-fraction described in the above, we will provide a simple but new method to obtain the mutual relation between the combined velocity correlation function $Z_\sigma^{\pm}(t)$ and $\gamma(t)$ in a short time region, in the following section.

## 6. Recursion formulae for $Z_\sigma^{\pm}(t)$ and $\gamma(t)$

Here, we provide a new and useful method to solve the Langevin equation based on recursion process [29]. Its detail is shown below.

Let us consider a Langevin equation for an evolution function being equivalent to (29) and (30), as follows:

$$dy(t)/dt = \int_0^t q(t-s)y(s)ds \tag{47}$$

The power expansion for $q(t)$ is defined as follows:

$$q(t) = \sum_{n=0}^{\infty} (q_n/n!)t^n \quad \left(q_n = q^{(n)}(0)\right) \tag{48}$$

and the corresponding expansion formula for $y(t)$ is written as follows:

$$y(t) = \sum_{m=0}^{\infty} (y_m/m!)t^m \quad \left(y_m = y^{(m)}(0)\right) \tag{49}$$

Putting (48) and (49) into the right hand side of Eq. (47), we have

$$\begin{aligned}\int_0^t q(t-s)y(s)ds &= \sum_{n,\,m=0}^{\infty} (q_n/n!)\,(y_m/m!)\int_0^t (t-s)^n s^m ds \\ &= \sum_{n,\,m=0}^{\infty} (q_n/n!)\,(y_m/m!)t^{(n+m+1)}\int_0^1 (1-p)^n p^m\, dp \\ &= \sum_{n,\,m=0}^{\infty} (q_n/n!)\,(y_m/m!)t^{(n+m+1)}B(n+1,m+1) \\ &= \sum_{n,\,m=0}^{\infty} (q_n/n!)\,(y_m/m!)t^{(n+m+1)}\{\Gamma(n+1)\Gamma(m+1)\}/\{\Gamma(n+m+2)\} \\ &= \sum_{n,\,m=0}^{\infty} (q_n/n!)\,(y_m/m!)t^{(n+m+1)}(n!m!)/(n+m+1)! \\ &= \sum_{n,\,m=0}^{\infty} \left(\left\{(q_n y_m)\,t^{(n+m+1)}/(n+m+1)!\right\} = \sum_{k=1}^{\infty}(z_k/k!)t^k\right.\end{aligned} \tag{50}$$

where B(n + 1, m + 1) and Γ(n + 1) mean the beta-function and the gamma-function, respectively, and

$$z_k = \sum_{k=n+m+1} q_n y_m \tag{51}$$

On the other hand, the left hand side of Eq. (47) is equal to the following formulae:

$$y'(t) = \left\{\sum_{k=0}^{\infty} (y_k/k!)t^k\right\}' = \sum_{k=0}^{\infty} (y_{k+1}/k!)t^k \tag{52}$$

Compare both expressions (50) and (52), we can get the recursion formulae as follows,

$$y_1 = 0; \quad y_{k+1} = \left\{\sum_{m=0}^{k-1} (q_{k-m-1} y_m)\right\} \quad (k = 1, 2, \dots) \tag{53}$$

Therefore, Eq. (49) is practically expressed in the following series:

$$\begin{aligned} &y_1 = 0; \quad y_2 = y_0 q_0; \quad y_3 = y_0 q_1 + y_1 q_0 = y_0\, q_1; \\ &y_4 = q_2\, y_0 + q_1\, y_1 + q_0\, y_2 = y_0\,(q_0{}^2 + q_2); \\ &y_5 = y_0 q_3 + y_1 q_2 + y_2 q_1 + y_3 q_0 = y_0\,(2q_0\, q_1 + q_3)\end{aligned} \tag{54}$$

and so on.

And vice versa, $q_n$'s are expressed as follows:

$$\begin{aligned} &q_1 = (1/y_0)y_3; \quad q_2 = (y_4/y_0) - q_0\,(y_2/y_0) = (1/y_0)\{y_4 - q_0 y_2)\}; \\ &q_3 = (1/y_0)\{y_5 - (y_2 y_3/y_0) - q_0\, y_3\}\end{aligned} \tag{55}$$

and so on.

This method can be immediately applicable in the following way, comparing with Eqs. (38) and (39).

$$q_0 = -\tilde{\gamma}(0)^2 \tag{56}$$

$$y(t) = y_0\left[1 - (t^2/2!)\tilde{\gamma}(0)^2 + \dots\right] \tag{57}$$

where

$$y_0 = (3k_BT/m^{\pm})\ (= Z_{\sigma}{}^{\pm}(0) \equiv Z_0{}^{\pm}) \tag{58}$$

## 7. Fluctuation dissipation theorem on the Laplace transformation of $\gamma(t)$

Considering Eqs. (56) and (57), the memory function $\gamma(t)$ can be taken as the following form:

$$\gamma(t) = \tilde{\gamma}(0)^2\, f(t) \tag{59}$$

where $f(0) = 1$.

The Laplace transformation of (59) in the long wavelength limit is then written as follows:

$$\tilde{\gamma}(0) = \tilde{\gamma}(0)^2 \int_0^{\infty} f(t)\, dt \tag{60}$$

Therefore, we have immediately,

$$\int_0^{\infty} f(t)\, dt = 1/\tilde{\gamma}(0) \tag{61}$$

On the other hand, the memory function and its Laplace transformation are described as in the following forms, by using the fluctuation dissipation theorem [6–9],

$$\gamma(t) = (1/3\mu k_B T) < \mathbf{R}_i(t)\, \mathbf{R}_j(0) > \tag{62}$$

and

$$\tilde{\gamma}(\omega) = (1/3\mu k_B T) \int_0^{\infty} \exp(-i\omega t) < \mathbf{R}_i(t)\, \mathbf{R}_j(0) > dt \tag{63}$$

The most simplest expression for $< \mathbf{R}_i(t)\ \mathbf{R}_j(0) >$ can be taken as in the following form:

$$< \mathbf{R}_i(t)\, \mathbf{R}_j(0) > \; = \; < \mathbf{R}_{ij}{}^2 > h(t) \tag{64}$$

where $< \mathbf{R}_{ij}{}^2 > = < \mathbf{R}_i(0)\ \mathbf{R}_j(0)>$.

Putting (64) into (62) and using (59), we have

$$\gamma(t) = (1/3\mu k_B T) < \mathbf{R}_{ij}{}^2 > h(t) = \tilde{\gamma}(0)^2 f(t) \tag{65}$$

This equation gives $h(t) \propto f(t)$, and if we take both functions are identical, then

$$< \mathbf{R}_{ij}{}^2 > \; = (1/3\mu k_B T)\tilde{\gamma}(0)^2 \tag{66}$$

Putting this relation into (62), we obtain again the relation (59), which indicates that the assumption, $h(t) = f(t)$, is exactly justified.

Therefore, the general form for the memory function $\gamma(t)$ is always written in the form of Eq. (59).

## 8. Former theories of velocity correlation functions in molten salts

Various analytic forms for memory functions were proposed [7, 8, 12, 39–43] and all these functions are qualitatively useful to obtain the combined velocity correlation functions, although some of these theories cannot predict the result obtained by MD simulation.

For example, if we use an approximate form for the memory function as

$$\begin{aligned} Z_\sigma^{\pm}(t) &= \left(3k_BT/m^{\pm}\right) \exp\left\{-\tilde{\gamma}(0)t/2\right\} \left[\cos\left(\sqrt{3}\tilde{\gamma}(0)t/2\right)\right. \\ &\quad \left. + \left\{\left(\tilde{\gamma}(0)/2\right)/\left(\sqrt{3}\tilde{\gamma}(0)/2\right)\right\} \sin\left(\sqrt{3}\tilde{\gamma}(0)t/2\right)\right] \\ &= \left(3k_BT/m^{\pm}\right)\left[1-\left(t^2/2!\right)\tilde{\gamma}(0)^2-\left(t^3/3!\right)3\tilde{\gamma}(0)^3/8+\left(\text{over}\, t^4\right)\right] \end{aligned} \tag{67}$$

As shown in our previous results [29], the calculated $Z_\sigma^{+}(t)$ for cation by using Eq. (67) agrees with that of MD simulation [7] qualitatively and semi-quantitatively.

However, the time expansion forms of $Z_\sigma^{\pm}(t)$ are essentially equal to the even powers expansion forms, which contradicts to the expression of (67). It is, therefore, necessary to seek an appropriate memory function which can be expanded as the even powers of the time t, even though the obtained result is numerically very close to the expression of $\gamma(t) = \tilde{\gamma}(0)^2 \exp\{-\tilde{\gamma}(0)t\}$.

## 9. Application of recursion method for the derivation of $\gamma(t)$ from $Z_\sigma^{\pm}(t)$

So far, we are successful to obtain the mutual relation between $\gamma(t)$ and $Z_\sigma^{\pm}(t)$ within a short time region to satisfy the Langevin equations in molten salts.

There are several works to obtain the auto-velocity correlation functions in monatomic liquids from appropriate memory functions $\gamma(t)$ [39, 41, 42].

However, it is not known what sorts of model functions are suitable for the combined velocity correlation function $Z_\sigma^{\pm}(t)$ until the present time. In order to elucidate this question, we will try to calculate the coefficients $y_m$'s of simulated $Z_\sigma^{\pm}(t)$ of molten NaCl in a short time region, and from these the corresponding $\gamma(t)$ will be obtained.

Previously we have already carried out the MD simulation for the combined velocity correlation functions $Z_\sigma^{\pm}(t)$ [7].

We try two types of power expansion forms in order to fit the combined correlation functions $Z_\sigma^{\pm}(t)$ by MD simulation. One is an arbitrary expansion form given by the even power series of the time $t$, which is theoretically exact for the combined correlation function. Another one is the series of even and odd powers for higher order terms over $t^2$ one. Practical reason for the use of latter case will be given below.

In the case of the utilization of only even powers, it was quite difficult to get to the simulated $Z_\sigma^{\pm}(t)$ even if the power's number is taken up to 36th order of time $t$.

On the other hand, we can get an agreement if we use even and odd serial powers over $t^2$ up to $t^9$. This fact encourages us that the combined velocity correlation functions $Z_\sigma^{\pm}(t)$ in molten systems must be practically analyzed in terms of even and odd powers of the time over $t^2$.

Therefore, the method utilizing the odd and even power series has a more rapid convergence for obtaining $Z_\sigma^{\pm}(t)$, in comparison with the method utilizing only even power series.

The fitting parameters, which are equal to $y_m$'s, are obtained by the non-linear least mean square method as so-called Levenberg-Marquart method [44].

The primary value in this non-linear least mean square method is inferred by utilization of simplex method.

It is inevitable that the coefficients of $y_m$'s (m = 3, 4, ... ) are slightly variable because of the termination effect in the expansion form. But we have no difficulty to elucidate $\gamma(t)$ in an appropriate short time range.

By using these obtained $y_m$'s, it is immediately possible to obtain $q_n$'s. And thereafter we can get a fitted curve indicating the curve of $\gamma(t)$ within a short time region. In this figure, the fitting curve of $\gamma(t)$ is obtained for the time range of $0 < t < 5.0 \times 10^{-14}$ seconds, from the expansion form of $Z_\sigma^{\pm}(t)$ up to $t^{15}$.

It is therefore emphasized that the utilization of odd terms within the short time region is necessary for the derivation of $q_n$'s from the $y_m$'s obtained by MD simulation.

For references, several analytic functional forms describing $\gamma(t)$ can also be given. The following two-types of functional forms are known as model functions being suitable for the auto-velocity correlation functions in liquids.

$$(\mathbf{a}-\mathbf{1})\quad \gamma(t) = \tilde{\gamma}(0)^2\,\mathrm{sech}\{(\pi/2)\tilde{\gamma}(0)t\} \tag{68}$$

$$(\mathbf{a}-2)\quad \gamma(t) = \tilde{\gamma}(0)^2\,\exp\left\{-(\pi/4)\tilde{\gamma}(0)^2t^2)\right\} \tag{69}$$

The $\gamma(t)$ is expressed by the form of $\tilde{\gamma}(0)^2\exp\{-\tilde{\gamma}(0)t\}$ agrees, at least within the short time region, with that of MD simulation.

However, an inevitable fact is that the theoretical memory function must be an expansion form of only even powers of the time, even though it is numerically close to the exponentially decaying function which includes the odd powers.

Is it possible to get a model function to fit the obtained curve of $\gamma(t)$ by MD simulation? To answer this question, we have carried out the fitting procedure by using a combination of poly-Gaussian functions [29]. Practically, the following form composed of three kinds of Gaussian functions is good enough to reproduce the obtained curve of $\gamma(t)$ under the condition of Eq. (61) for molten NaCl at 1100 K,

$$\gamma(t) = \tilde{\gamma}(0)^2\sum_{i=1}^{3} a_i \exp\left\{-(\pi/4)\, b_i\tilde{\gamma}(0)^2t^2)\right\} \tag{70}$$

where

$$\sum_{i=1}^{3} a_i = 1, \quad \text{and } \left\{(b_2b_3)^{1/2}a_1 + (b_3b_1)^{1/2}a_2 + (b_1b_2)^{1/2}a_3\right\}/(b_1b_2b_3)^{1/2} = 1 \tag{71}$$

Using (70) and (71), we could reproduce the obtained curve of $\gamma(t)$ by MD simulation in molten NaCl at 1100 K. And these are approximated to as $\{a_1 = 0.2, a_2 = 0.3 \text{ and } a_3 = 0.5\}$, which values correspond to the existing fractions of each short range configuration i = 1, i = 2, and i = 3, respectively. And values of $\{b_1 = 97.50, b_2 = 6.52, \text{ and } b_3 = 0.38\}$ correspond to their structural decaying speeds, respectively.

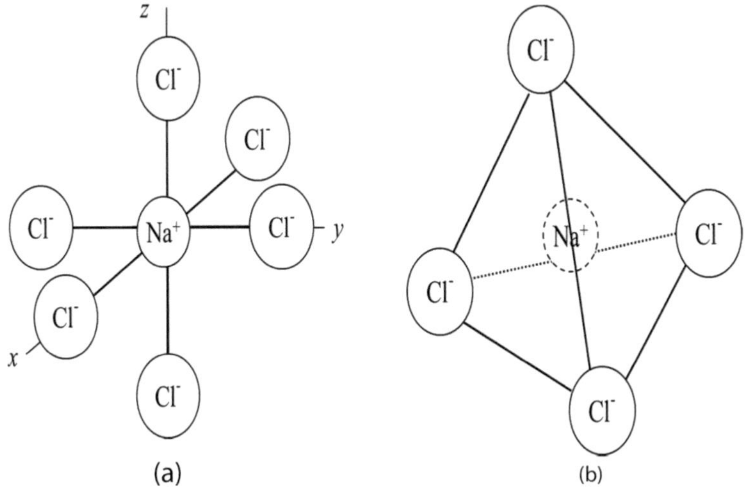


**Figure 3.**
*(a) A stable short range configuration of 6 $Cl^-$ ions around a $Na^+$ ion. (b) Another stable short range configuration of 4 $Cl^-$ ions around a $Na^+$ ion.*

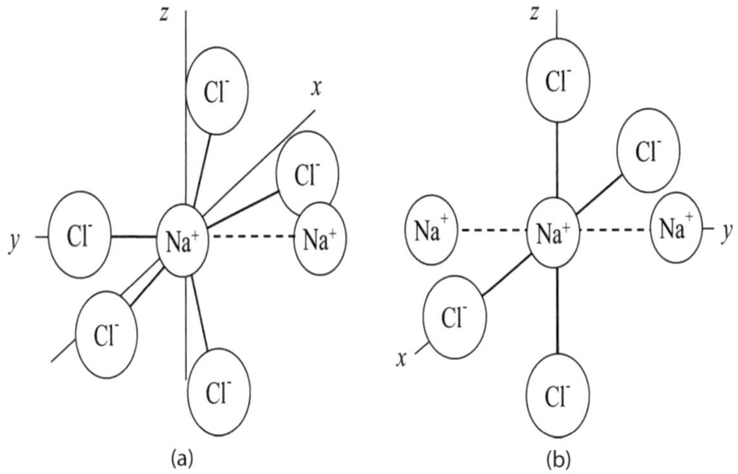


**Figure 4.**
*(a) A rather unstable short range configuration of 5 $Cl^-$ ions around a $Na^+$ ion. (b) Another unstable short range configuration of 4 $Cl^-$ ions around a $Na^+$ ion.*

| Degree of stability | Configuration type | | | |
|---|---|---|---|---|
| | **Coordination of 4 $Cl^-$ ions** | **Coordination of 5 $Cl^-$ ions** | **Coordination of 6 $Cl^-$ ions** | **Existing probability, $a_i$** |
| i = 1 | 0.2 | | | 0.2 |
| i = 2 | | 0.3 | | 0.3 |
| i = 3 | 0.15 | | 0.35 | 0.5 |

**Table 1.**
*Local configuration types of $Cl^-$ ions around a centered $Na^+$ ion.*

According to **Figure 2a**, the averaged nearest neighbor's number around the $Na^+$ ion is equal to 5.0. Any local coordination numbers around a $Na^+$ are possible to be 4, 5, and 6 under the condition of density fluctuation in the liquid state.

It is possible to consider that stable short range configurations seem to be two types. One is the case of cubic structure-type configuration having with the coordination of 6 chlorine ions around the centered sodium ion as shown in **Figure 3a**, which is similar to the solid type configuration with a sort of lengthen fluctuation of the interionic distance.

The other is close to a tetrahedral coordination of chlorine ions around the centered sodium ion as shown in **Figure 3b**.

For simplicity, here we assume that the decaying or releasing of these two types of rather stable short range configurations is nearly the same, then the combined configurational decaying is given by i = 3 and $b_3$.

On the other hand, there exist two types of rather unstable short range configurations as shown in **Figure 4a** and **b**, respectively, in which the surrounded $Cl^-$ ions around a $Na^+$ ion are spatially asymmetric.

Totally, the local configuration types of $Cl^-$ ions around a centered $Na^+$ ion are listed in **Table 1**.

## 10. Discussion and conclusions in the case of molten salts

As shown in the previous section, the combined velocity correlation functions $Z_\sigma^{\pm}(t)$ can be analyzed in terms of odd and even powers over $t^2$ in their expansion forms and the corresponding memory function includes the terms of odd and even powers in its expansion form.

In addition, it is emphasized that the $\gamma(t)$ obtained from the simulated $Z_\sigma^{+}(t)$ agrees completely with that from $Z_\sigma^{-}(t)$. This fact means that the memory functions for cation and anion are identical and Eq. (15) is automatically justified by the present new type of experiment such as computer simulation.

In conclusion, we have newly obtained the mutual relation between the memory function $\gamma(t)$ and the combined velocity correlation function $Z_\sigma^{\pm}(t)$, by using a recursion method to solve the Langevin equation and it may be applicable for finding a suitable memory function in all liquid matters.

## 11. Generalized Langevin equation in electrolytic solution

Hereafter, we will consider the strong electrolytic solution composed of $N^+$ cations, $N^-$ anions and X water molecules in a volume $V_M$. For simplicity, we take that $N^+ = N^- = N$ and ions charges are equal to $z^+ = -z^- = z$. Then the number densities of ions and water molecules are equal to $n^+ = n^- = n = N/V_M$ and $x = X/V_M$, respectively. And furthermore we assume that the dissociation of electrolyte is complete under the condition of $N \ll X$.

In the present system, a generalized Langevin equation for the cation (or anion) i under an external field **E** is written as follows:

$$m^{\pm}d\mathbf{v}_i^{\pm}(t)/dt = -m^{\pm}\int_0^t \gamma^{\pm}(t-t')\mathbf{v}_i^{\pm}(t')dt' + z^{\pm}e(\mathbf{E}+\mathbf{F}) + z^{\pm}e\varepsilon_i(t) \qquad (72)$$

where $\gamma^{\pm}(t)$ is the memory function incorporating with the friction force acting on its cation (or anion). **F** is the induced internal field yielded by the change of ion's distribution which is resulted from the applying external field **E**, and $\varepsilon_i(t)$ is the random fluctuating force acting on the ion i.

According to Berne and Rice [16], the internal field **F** induced by the asymmetric ion's distribution in an ionic melt is expressed as follows:

$$\mathbf{F} = -\delta \cdot \mathbf{E} = -4\pi n/3k_BT\int_d^\infty \{d\phi^{+-}(r)/dr\}g^{+-}(r)r^3dr \cdot \mathbf{E} \tag{73}$$

where $g^{+-}(r)$ is the pair distribution function between cation and anion, and d is the hard-core contact distance between cation and anion. Hereafter, we will use this result.

If we take $\mathbf{E} = \mathbf{E}_0\, e^{-i\omega t}$, then the ensemble average for $\mathbf{v}_i^{\pm}(t)$ is written in the following form:

$$<\mathbf{v}_i^{\pm}(t)> \;=\; \mathrm{Re}\ \mu^{\pm}(\omega)z^{\pm}e\,\mathbf{E}_0\, e^{-i\omega t} \tag{74}$$

Inserting (74) into (72) and taking ensemble average under the assumption of $<\varepsilon_i(t)> = 0$, we have

$$m^{\pm} < d\mathbf{v}_i^{\pm}(t)>/dt = -m^{\pm}\int_0^t \gamma^{\pm}(t-t')\,\mathbf{v}_i^{\pm}(t')dt' + z^{\pm}e(1\text{–}\delta)\mathbf{E} \tag{75}$$

Therefore,

$$\mu^{\pm}(\omega) = (1\text{–}\delta)/m^{\pm}\{-i\omega + \tilde{\gamma}^{\pm}(\omega)\} \tag{76}$$

where

$$\tilde{\gamma}^{\pm}(\omega) = \int_0^\infty \gamma^{\pm}(t)\, e^{i\omega t}\, dt \tag{77}$$

The dc current density $\mathbf{j}^{\pm}$ is then written as follows:

$$\mathbf{j}^{\pm} = nz^{\pm}e<\mathbf{v}_i^{\pm}(t)>_{\omega=0} = nz^2e^2\mu^{\pm}(0)\,\mathbf{E}_0 = nz^2e^2(1\text{–}\delta)\,\mathbf{E}_0/m^{\pm}\tilde{\gamma}^{\pm}(0) \tag{78}$$

On the other hand, $\mathbf{j}^{\pm}$ is expressed as $\mathbf{j}^{\pm} = \sigma^{\pm}\mathbf{E}_0$, where $\sigma^{\pm}$ being equal to the partial conductivity for cation or anion. Therefore, $\sigma^{\pm}$ is written as follows:

$$\sigma^{\pm} = nz^2e^2(1\text{–}\delta)/m^{\pm}\tilde{\gamma}^{\pm}(0) \tag{79}$$

The Laplace transformation of the memory function in the long wavelength limit $\tilde{\gamma}^{\pm}(0)$ in Eq. (79) will be obtained in later section.

In the next section, we will discuss velocity correlation functions.

## 12. Linear response theory for electrolytic solutions

Eq. (79) is also obtainable from the following simplified Langevin equation:

$$m^{\pm}d\mathbf{v}_i^{\pm}(t)/dt = -m^{\pm}\tilde{\gamma}^{\pm}(0)\mathbf{v}_i^{\pm}(t) + z^{\pm}e(\mathbf{E}+\mathbf{F}) + z^{\pm}e\varepsilon_i(t) \tag{80}$$

Its derivation can be easily seen in a standard book of statistical physics.

Starting from Eq. (80) with an infinitesimal external field **E**, it is also easily obtainable the following Kubo-Green formulae for the partial conductivities $\sigma^+$ and $\sigma^-$ [6, 7, 28]:

$$\sigma^+ = (1/3k_BT)\int_0^\infty <\mathbf{j}^+(t)\cdot\mathbf{j}(0)> dt \tag{81}$$

and

$$\sigma^- = (1/3k_BT)\int_0^\infty <\mathbf{j}^-(t)\cdot\mathbf{j}(0)> dt \tag{82}$$

where the current densities $\mathbf{j}^{\pm}(t)$ and $\mathbf{j}(t)$ are defined by the following expressions:

$$\mathbf{j}^{\pm}(t) = z^{\pm}e\sum_n \mathbf{v}_i^{\pm}(t) \quad \text{and} \quad \mathbf{j}(t) = \mathbf{j}^+(t) + \mathbf{j}^-(t) \tag{83}$$

In order to obtain the partial conductivities based on Eqs. (81) and (82), it is necessary to study the velocity correlation functions, $< \mathbf{v}_i^+(t)\, \mathbf{v}_j^+(0)>$, $< \mathbf{v}_k^-(t)\, \mathbf{v}_l^-(0) >$ and $< \mathbf{v}_i^+(t)\, \mathbf{v}_k^-(0)>$.

In the next section, we will discuss velocity correlation functions described in terms of inter-molecular (or ionic) potentials and pair distribution functions in order to obtain the $\tilde{\gamma}^{\pm}(0)$.

## 13. Short time expansion of velocity correlation functions in electrolytic solutions

The short time expansion of velocity correlation function, $< \mathbf{v}_i^+(t)\, \mathbf{v}_j^+(0) >$ for cation is written as

$$\begin{aligned} <\mathbf{v}_i^+(t)\,\mathbf{v}_j^+(0)> &= <\mathbf{v}_i^+(0)\,\mathbf{v}_j^+(0)> + t^2/2!<\mathbf{v}_i^+(0)\,\ddot{\mathbf{v}}_j^+(0)> \\ &\quad + \left(\text{higher order over} t^4\right) \end{aligned} \tag{84}$$

In the present aqueous solution of electrolyte, the total Hamiltonian of the system is written as follows:

$$H = \sum_{i=1}^{N+} \mathbf{p}_i^{+2}/2m^+ + \sum_{k=1}^{N-} \mathbf{p}_i^{-2}/2m^- + \sum_{q=1}^{X} \mathbf{p}_q^{w2}/2m^w + V \tag{85}$$

where

$$\begin{aligned} V = &\sum_{i\neq j}^{N+} \phi^{++}\left(|\mathbf{r}_i^+ - \mathbf{r}_j^+|\right) + \sum_{k\neq l}^{N-} \phi^{--}\left(|\mathbf{r}_k^- - \mathbf{r}_l^-|\right) + \sum_{i,k}^{N+,N-} \phi^{+-}\left(|\mathbf{r}_i^+ - \mathbf{r}_k^-|\right) \\ &+ \sum_{i,q}^{N+,X} \phi^{+w}\left(|\mathbf{r}_i^+ - \mathbf{r}_q^w|\right) + \sum_{k,q}^{N-,X} \phi^{-w}\left(|\mathbf{r}_k^- - \mathbf{r}_q^w|\right) + \sum_{q,s}^{X} \phi^{ww}\left(|\mathbf{r}_q^w - \mathbf{r}_s^w|\right) \end{aligned} \tag{86}$$

Since the water molecule is not spherical in its molecular configuration, it is difficult to define the position of $\mathbf{r}_q^w$. However, we tentatively assume that its position is located at the center of oxygen atom in the $H_2O$ molecule.

From the Poisson's equation of motion,

$$\mathbf{p}_i^+ \ddot{\mathbf{p}}_i^+ = -\sum_{j=1}^{N+} \left(\mathbf{p}_i{}^+ \mathbf{p}_j{}^+ / m^+\right) \partial^2 V / \partial \mathbf{r}_i{}^+ \partial \mathbf{r}_j{}^+ - \sum_{k=1}^{N-} \left(\mathbf{p}_i{}^+ \mathbf{p}_k{}^- / m^-\right) \partial^2 V / \partial \mathbf{r}_k{}^- \partial \mathbf{r}_i{}^+$$

$$- \sum_{q=1}^{X} \left(\mathbf{p}_i{}^+ \mathbf{p}_q{}^w / m^w\right) \partial^2 V / \partial \mathbf{r}_q{}^w \partial \mathbf{r}_i{}^+ \qquad (87)$$

and

$$\mathbf{p}_i^+ \ddot{\mathbf{p}}_{i \neq i'}^+ = -\sum_{j}^{N+} \left(\mathbf{p}_{i'}{}^+ \mathbf{p}_j{}^+ / m^+\right) \partial^2 V / \partial \mathbf{r}_i{}^+ \partial \mathbf{r}_j{}^+ - \sum_{k=1}^{N-} \left(\mathbf{p}_i{}^+ \mathbf{p}_k{}^- / m^-\right) \partial^2 V / \partial \mathbf{r}_k{}^- \partial \mathbf{r}_i{}^+$$

$$- \sum_{q=1}^{X} \left(\mathbf{p}_i{}^+ \mathbf{p}_q{}^w / m^w\right) \partial^2 V / \partial \mathbf{r}_q{}^w \partial \mathbf{r}_i{}^+ \qquad (88)$$

Since the second derivative of the potential term V is independent for the product of momenta, all other terms other than i = j in (87) must vanish on averaging. And in a similar way, the meaningful terms of (88) for averaging must be also equal to the case i ≠ i' = j. Therefore, taking the ensemble averages for (87) and (88), we have

$$< \mathbf{p}_i^+ \ddot{\mathbf{p}}_j^+ > = < \mathbf{p}_i^+ \ddot{\mathbf{p}}_i^+ > + < \mathbf{p}_i^+ \ddot{\mathbf{p}}_{j \neq i}^+ > = -k_B T \{ n < \phi^{+-} > + x < \phi^{+w} > \} \qquad (89)$$

where

$$< \phi^{+-} > = \int_0^\infty \{ \partial^2 \phi^{+-}(r) \partial r^2 + (2/r) \partial \phi^{+-}(r) \partial r \} g^{+-}(r) \, 4\pi r^2 dr \qquad (90)$$

and

$$< \phi^{+w} > = \int_0^\infty \{ \partial^2 \phi^{+w}(r) \partial r^2 + (2/r) \partial \phi^{+w}(r) \partial r \} g^{+w}(r) \, 4\pi r^2 dr \qquad (91)$$

In this equation, $g^{+w}(r)$ is the pair distribution function between cation and water molecule.

It is emphasized that there is no contribution from $\phi^{++}(r)$ in Eq. (89) because of the cancelation by the terms of i = j and i ≠ j in $< \mathbf{p}_i^+ \ddot{\mathbf{p}}_j^+ >$ [7].

Insertion of (89) into (84) gives us the following form:

$$Z^+(t) \equiv < \mathbf{v}_i{}^+(t) \, \mathbf{v}_j{}^+(0) >$$

$$= < \mathbf{v}_i{}^+(0) \, \mathbf{v}_j{}^+(0) > - \left(t^2/2!\right) k_B T \{ (n < \phi^{+-} > + x < \phi^{+w} >) / m^+ m^+ \}$$

$$+ \left(\text{higher order over} \, t^4\right) \qquad (92)$$

In a similar way, the term $< \mathbf{p}_i^+ \ddot{\mathbf{p}}_k^- >$ can be described as follows:

$$< \mathbf{p}_i^+ \ddot{\mathbf{p}}_k^- > = k_B T \, n < \phi^{+-} > \qquad (93)$$

Using this relation, the distinct velocity correlation function is written as follows:

$$\begin{aligned} <\mathbf{v}_i^+(t)\,\mathbf{v}_k^-(0)> &= <\mathbf{v}_i^+(0)\,\mathbf{v}_k^-(0)> + t^2/2! <\mathbf{v}_i^+(0)\,\ddot{\mathbf{v}}_k^-(0)> \\ &\quad + (\text{higher order over}\, t^4) \\ &= <\mathbf{v}_i^+(0)\,\mathbf{v}_k^-(0)> + (t^2/2!)\mathrm{k_B T n} <\phi^{+-}>/\mathrm{m^+ m^-} \\ &\quad + (\text{higher order over}\, t^4) \end{aligned} \tag{94}$$

Using (92) and (94), the combined velocity correlation function $Z_\sigma^+(t)(= <\mathbf{j}^+(t)\,\mathbf{j}(0)>/n^2z^2e^2)$ incorporation with the partial conductivity $\sigma^+$ is therefore expressed as follows:

$$\begin{aligned} Z_\sigma^+(t) &\equiv <\mathbf{v}_i^+(t)\,\mathbf{v}_j^+(0)> - <\mathbf{v}_i^+(t)\,\mathbf{v}_k^-(0)> \\ &= (3\mathrm{k_B T/m^+})\{1-(t^2/2!)(\mathrm{n}<\phi^{+-}>/3\mu + \mathrm{x}<\phi^{+w}>/3\mathrm{m}^+) \\ &\quad +(\text{higher order over}\, t^4)\} \end{aligned} \tag{95}$$

where $\mu$ is equal to the reduced mass of $m^+$ and $m^-$. In deriving (95), we have assumed the initial conditions as follows:

$$<\mathbf{v}_i^+(0)\,\mathbf{v}_j^+(0)> = <\mathbf{v}_i^+(0)\,\mathbf{v}_i^+(0)> = (3\mathrm{k_B T/m^+}) \quad \text{and} \quad <\mathbf{v}_i^+(0)\,\mathbf{v}_k^-(0)> = 0 \tag{96}$$

These initial conditions are confirmed by our own molecular dynamic simulation, which will be shown in the later section. In a similar way, we have

$$\begin{aligned} Z_\sigma^-(t) &\equiv <\mathbf{v}_i^-(t)\,\mathbf{v}_j^-(0)> - <\mathbf{v}_i^+(0)\,\mathbf{v}_j^-(0)> \\ &= (3\mathrm{k_B T/m^-})\{1-(t^2/2!)\,(\mathrm{n}<\phi^{+-}>/3\mu + \mathrm{x}<\phi^{-w}>/3\mathrm{m}^-) \\ &\quad +(\text{higher order over}\, t^4)\} \end{aligned} \tag{97}$$

where

$$<\phi^{-w}> = \int_0^\infty \{\partial^2\phi^{-w}(r)\partial r^2 + (2/r)\partial\phi^{-w}(r)\partial r\}\mathrm{g}^{-w}(r)\,4\pi r^2\,\mathrm{d}r \tag{98}$$

$g^{-w}(r)$ of this equation means the pair distribution function between anion and water molecule. And it is also emphasized that the contribution from $\phi^{--}(r)$ to $<\mathbf{v}_i^-(t)\,\mathbf{v}_j^-(0)>$ is also vanished to be zero.

It is impossible to obtain the partial conductivities by the insertion of (95) and (97) into (81) and (82), because we knew only the terms up to $t^2$ in their explicit forms. However, these equations can be utilized for the derivation of $\tilde{\gamma}^{\pm}(0)$ as shown in the next section.

## 14. Derivation of $\tilde{\gamma}^{\pm}(0)$ in electrolytic solutions

According to the fluctuation dissipation theorem applied for the present system with the condition of no external field or of infinitesimal external field, the Laplace transformation of the memory function $\gamma^{\pm}(t)$ and that of the ensemble average of time correlation function for the fluctuating random force $<\varepsilon_i(t)\varepsilon_i(0)>$ have the following relation [25, 28]:

$$m^{\pm 2} < \mathbf{v}_i^{\pm}(0) \cdot \mathbf{v}_j^{\pm}(0) > \tilde{\gamma}^{\pm}(\omega) = m^{\pm 2} < \mathbf{v}_i^{\pm}(0) \cdot \mathbf{v}_i^{\pm}(0) > \tilde{\gamma}^{\pm}(\omega)$$
$$= z^2 e^2 \int_0^{\infty} < \varepsilon_i(t)\varepsilon_i(0) > e^{i\omega t}\, dt \tag{99}$$

The fluctuation dissipation theorem tells us the following relation:

$$\tilde{\gamma}^{\pm}(\omega) = \left(1/3\, m^{\pm} k_B T\right)\left(z^2 e^2\right) \int_0^{\infty} < \varepsilon_i(t)\varepsilon_i(0) > e^{i\omega t}\, dt \tag{100}$$

In the long wavelength limit, this relation is expressed by

$$\tilde{\gamma}^{\pm}(0) = \left(1/3\, m^{\pm} k_B T\right) \left(z^2 e^2\right) \int_0^{\infty} < \varepsilon_i(t)\varepsilon_i(0) > dt \tag{101}$$

Let us go back to the memory function $\gamma^{\pm}(t)$ and assume a combined exponential decay functions for it as follows, although this assumption is not exactly consistent with Eq. (84), but technically acceptable one as discussed in the case of molten salt [29],

$$\gamma^{\pm}(t) = \gamma_0^{\pm}(t)(\{= \gamma_{00}^{\pm} \exp\left(-\beta_0^{\pm} t\right)\} + \gamma_1^{\pm}(t)\{= \gamma_{01}^{\pm} \exp\left(-\beta_1^{\pm} t\right)\} \tag{102}$$

In this equation, the pre-exponential factor $\gamma_{00}^{\pm}$ is subject to the interactions between the central ion and surrounding water molecules. The decaying constants are related to the time dependence of its configuration change. The pre-exponential factor, $\gamma_{01}^{\pm}$, is equal to the magnitude of memory function at t = 0 in respect to the friction force acting on the central cation or anion caused by interactions between its central ion and neighboring ions. In other words, the first term on the right hand side of this equation means the case of dilute limit of electrolytic solution and the second one is equal to the effective friction caused by the addition of a moderate amount of electrolyte. Therefore, the first term is related to either $<\phi^{+w}>$ or $<\phi^{-w}>$, while the second one is related to the term $<\phi^{+-}>$.

Using (94) and (96), $\gamma_{00}^{\pm}$ and $\gamma_{01}^{\pm}$ are expressed as follows:

$$\gamma_{00}^{\pm} + \gamma_{01}^{\pm} = -\ddot{Z}_{\sigma}^{\pm}(0)/Z_{\sigma}^{\pm}(0) = x < \phi^{\pm w} > /3m^{\pm} + n < \phi^{+-} > /3\mu \tag{103}$$

In the dilute limit of $n \ll x$, we have

$$\left(\gamma_{00}^{\pm}\right) = x < \phi^{\pm w} > /3m^{\pm} \tag{104}$$

And then we have

$$\left(\gamma_{01}^{\pm}\right) = n < \phi^{+-} > /3\mu \tag{105}$$

At the dilution limit of electrolyte where the contribution of $\gamma_1^{\pm}(t)$ can be neglected, the Laplace transformation of $\gamma_0^{\pm}(t)$ in the long wavelength limit is then described as follows:

$$\tilde{\gamma}_0^{\pm}(0) = \gamma_{00}^{\pm}/\beta_0^{\pm} = x < \phi^{\pm w} > /3m^{\pm}\beta_0^{\pm}$$
$$= \left(1/3\, m^{\pm} k_B T\right) \left(z^2 e^2\right) \int_0^{\infty} < \varepsilon_{i0}(t)\varepsilon_{i0}(0) > dt \tag{106}$$

where the auto-correlation function of random fluctuating force $<\varepsilon_{i0}(t)\varepsilon_{i0}(0)>$ is only related to either $<\phi^{+w}>$ or $<\phi^{-w}>$.

As seen in Eq. (79), the Laplace transformation of memory function in the long wavelength limit, $\tilde{\gamma}_0^{\pm}(0)$, corresponds to effective friction constants for cation and anion, which means that the auto-correlation function of random fluctuating force.

$<\varepsilon_{i0}(t)\ \varepsilon_{i0}(0)>$ may be represented by an exponential decaying function with the time constant of $\tilde{\gamma}_0^{\pm}(0)$ as follows:

$$<\varepsilon_{i0}(t)\varepsilon_{i0}(0)> \; = \; <\varepsilon_{i0}(0)\varepsilon_{i0}(0)> \exp\left\{-\tilde{\gamma}_0^{\pm}(0)t\right\} \tag{107}$$

Insertion of (107) into (106) gives us

$$<\varepsilon_{i0}(0)\varepsilon_{i0}(0)> \; = 3\,\mathrm{m}^{\pm}\mathrm{k_B T}\left(\tilde{\gamma}_0^{\pm}(0)\right)^2 \tag{108}$$

Therefore, we obtain

$$\gamma_0^{\pm}(t) = \left(\tilde{\gamma}_0^{\pm}(0)\right)^2 \exp\left(-\tilde{\gamma}_0^{\pm}(0)t\right) \tag{109}$$

Compare (106) and (109) we have

$$\beta_0{}^{\pm} = \tilde{\gamma}_0^{\pm}(0) = \left(\gamma_{00}{}^{\pm}\right)^{1/2} = \left(\mathrm{x} < \phi^{\pm \mathrm{w}} > /3\mathrm{m}^{\pm}\right)^{1/2} \tag{110}$$

By the analogy with this relation, we can infer the following relation:

$$\beta_1{}^{\pm} = \tilde{\gamma}_1^{\pm}(0) = \left(\gamma_{01}{}^{\pm}\right)^{1/2} = \left(\mathrm{n} < \phi^{\pm \mathrm{w}} > /3\mu\right)^{1/2} \tag{111}$$

Therefore, Eq. (102) is explicitly written as follows:

$$\begin{aligned}\gamma^{\pm}(t) &= \gamma_0{}^{\pm}(t) + \gamma_1{}^{\pm}(t)\\ &= \left(\mathrm{x} < \phi^{\pm \mathrm{w}} > /3\mathrm{m}^{\pm}\right) \exp\left[-\left(\mathrm{x} < \phi^{\pm \mathrm{w}} > /3\mathrm{m}^{\pm}\right)^{1/2} t\right]\\ &\quad + \left(\mathrm{n} < \phi^{+-} > /3\mu\right) \exp\left[-\left(\mathrm{n} < \phi^{+-} > /3\mu\right)^{1/2} t\right]\end{aligned} \tag{112}$$

And the Laplace transformation of this equation in the long wavelength limit is equal to

$$\tilde{\gamma}^{\pm}(0) = \left[\left(\mathrm{x} < \phi^{\pm \mathrm{w}} > /3\mathrm{m}^{\pm}\right)^{1/2}\right] + \left[\left(\mathrm{n} < \phi^{+-} > /3\mu\right)^{1/2}\right] \tag{113}$$

## 15. Partial conductivities $\sigma^+$ and $\sigma^-$

Putting Eq. (113) into (79), we obtain the formulae of the partial conductivities, $\sigma^+$ and $\sigma^-$, which are expressed in terms of the pair distribution functions and pair potentials as follows [28],

$$\begin{aligned}\sigma^{+} &= \mathrm{nz^2e^2}(1\text{–}\delta)/\mathrm{m}^{+}\tilde{\gamma}^{+}(0)\\ &= \mathrm{nz^2e^2}(1\text{–}\delta)/\mathrm{m}^{+}\left[\left(\mathrm{x} < \phi^{+\mathrm{w}} > /3\mathrm{m}^{+}\right)^{1/2} + \left(\mathrm{n} < \phi^{+-} > /3\mu\right)^{1/2}\right]\end{aligned} \tag{114}$$

and

$$\begin{aligned}\sigma^{-} &= \mathrm{nz^2e^2}(1\text{–}\delta)/\mathrm{m}^{-}\tilde{\gamma}^{-}(0)\\ &= \mathrm{nz^2e^2}(1\text{–}\delta)/\mathrm{m}^{-}\left[\left(\mathrm{x} < \phi^{-\mathrm{w}} > /3\mathrm{m}^{-}\right)^{1/2} + \left(\mathrm{n} < \phi^{+-} > /3\mu\right)^{1/2}\right]\end{aligned} \tag{115}$$

If the concentration c is defined as the number of moles of electrolyte in the unit volume (actually taken as 1 cc), then the number density n is equal to $cN_A$, where $N_A$ being the Avogadro's number. Then, the partial conductivities, $\sigma^+$ and $\sigma^-$, are written as follows:

$$\sigma^+ = ne\mu^+ = cN_A z^2e^2(1–\delta)/m^+\left[(x<\phi^{+w}>/3m^+)^{1/2} + (cN_A<\phi^{+-}>/3\mu)^{1/2}\right] \tag{116}$$

and

$$\sigma^- = ne\mu^- = cN_A z^2e^2(1–\delta)/m^-\left[(x<\phi^{-w}>/3m^-)^{1/2} + (cN_A<\phi^{+-}>/3\mu)^{1/2}\right] \tag{117}$$

In these equations, $\mu^+$ and $\mu^-$ are called as the mobility of cation and anion.

The partial molar conductance $\Lambda^+$ and $\Lambda^-$ are defined as $\Lambda^\pm = \sigma^\pm/c$. Then the total conductance $\Lambda_c$ is written as follows:

$$\Lambda_c = \Lambda^+ + \Lambda^- = N_A z^2e^2(1–\delta) \times \left[1/m^+\left\{(x<\phi^{+w}>/3m^+)^{1/2} + (cN_A<\phi^{+-}>/3\mu)^{1/2}\right\} + 1/m^-\left\{(x<\phi^{-w}>/3m^-)^{1/2} + (cN_A<\phi^{+-}>/3\mu)^{1/2}\right\}\right] \tag{118}$$

Under the condition of $n(=cN_A) \ll x$, they are approximated to as follows:

$$\Lambda^+ = N_A z^2e^2(1–\delta)(3/xm^+<\phi^{+w}>)^{1/2}\left[1–(cN_A m^+<\phi^{+-}>/\mu x<\phi^{+w}>)^{1/2}\right] \tag{119}$$

and

$$\Lambda^- = N_A z^2e^2(1–\delta)(3/xm^-<\phi^{-w}>)^{1/2}\left[1–(cN_A m^-<\phi^{+-}>/\mu x<\phi^{-w}>)^{1/2}\right] \tag{120}$$

From Eqs. (119) and (120), we have a form of $\Lambda_c = (\Lambda^+ + \Lambda^-) \simeq \Lambda_0 + \Lambda_1 – kc^{1/2}$. $\Lambda_0$ and k are written as follows:

$$\Lambda_0 = N_A z^2e^2\left\{(3/xm^+<\phi^{+w}>)^{1/2} + (3/xm^-<\phi^{-w}>)^{1/2}\right\} \tag{121}$$

$$\Lambda_1 = N_A\delta z^2e^2\left\{(3/xm^+<\phi^{+w}>)^{1/2} + (3/xm^-<\phi^{-w}>)^{1/2}\right\} \tag{122}$$

and

$$k = N_A z^2e^2(1–\delta)(3N_A<\phi^{+-}>/\mu)^{1/2}\{(1/x<\phi^{+w}> + (1/x<\phi^{-w}>)\} \tag{123}$$

As seen in these expressions, $\Lambda_0$ means the conductance in the dilution limit of electrolyte and $\Lambda_1$ is the correction term appeared by the so-called relaxation effect. The last term $kc^{1/2}$ is composed of the so-called electrophoretic effect and the combined term of both effects.

In the case of a moderate concentration of electrolyte, in particular, of relatively weak electrolyte, we have to take account of the degree of association between the opposite ions into the expression for the partial conductivities.

## 16. Pair potentials in electrolytic solution

A number of research works to obtain the model potentials in electrolytic solutions have been presented since the Debye-Hückel theory [18]. In particular, various qualified model potentials, which satisfy the experimental data such as the hydration free energy and the enthalpies in condensed and gas phases, have recently been proposed in order to carry out the molecular dynamic simulation. It is not our intension to compare or evaluate these potentials and therefore we like to refer only some of these for our interests [24–27, 45]. It may be possible to estimate these potentials by using wave mechanical approach. In fact the ion-water molecule interactions were obtained by such an elaborating method [46–48].

The essential point for these model potentials in electrolytic solutions is that the dielectric character should be concerned. According to Sack [49], the water-molecules around the constituent ion are strongly oriented and the ion's orientating ability to neighboring water-molecules decreases with increasing of the distance between the ion and those water-molecules. Oka [50] also estimated the change of effective dielectric constant as a function of distance between the ion and water-molecule.

We propose the following model function to satisfy these results as follows:

$$\varepsilon(r) = 1 + (\varepsilon_0 - 1)[1 - \exp\{-(r - \mathrm{r}_0)/\kappa\}] \tag{124}$$

where $\varepsilon_0$ (=78.35) is the dielectric constant of water. Other parameters are numerically equal to $\mathrm{r}_0$ = 5 A and $\kappa$ = 3.44 $A^{-1}$, respectively.

The insertion of this dielectric function $\varepsilon(r)$ for the long range part of inter-particle potential is not necessary in molecular dynamic simulations. The dynamics produces automatically the configuration of constituents to satisfy the dielectric behavior.

On the analogy of the inter-ionic potentials in molten salts, $\phi^{+-}(r)$ in aqueous solution, where the dipole-dipole and dipole-quadrupole dispersion forces are neglected, may be given as follows:

$$\phi^{+-}(r) = \left(\mathrm{z}^{+}\mathrm{z}^{-}\mathrm{e}^{2}\right)/\{r\,\varepsilon(r)\} + \mathrm{A}^{+-}\exp\left[\mathrm{B}^{+-}\left(\mathrm{d_i}^{+} + \mathrm{d_j}^{-} - r\right)\right] \tag{125}$$

where $A^{+-}$ is a constant in relation to the magnitude of repulsive force between cation i and anion j. $B^{+-}$ the softness parameter and ($d_i^{+}$ + $d_j^{-}$) is the hard core contact between cation i and anion j. $A^{+-}$ and $B^{+-}$ are also given in the literature [27]. The difference between this expression and that of ionic crystal or of molten salt is only ascribed to whether the introduction of the dielectric function $\varepsilon(r)$ is necessary or not.

For simplicity, the pair potentials $\phi^{+\,w}(r)$ and $\phi^{-\,w}(r)$ are assumed to be a combined form of the repulsive potential and the charge-dipole potential. Then the pair potential between cation and water molecule centered at the oxygen atom, $\phi^{+\,w}(r)$, is written as follows:

$$\phi^{+w}(r) = \phi_{\mathrm{rep}}{}^{+w}(r) - \mathrm{z}^{+}\mathrm{e}\mu\cos\theta\left(1\text{–}3\mathrm{l}^{2}/8r^{2}\right)/\left\{r^{2}\varepsilon(r)\right\} \tag{126}$$

where $\phi_{rep}{}^{+w}(r)$ is repulsive potential and its explicit form will be given be later. $\mu$ is the dipole of water molecule and l its length. $\theta$ is the dipole's angle in the direction of cation.

It is well-known that the above expression is converted to the following form according to Boltzmann law,

$$\phi^{+w}(r) = \phi_{rep}{}^{+w}(r) - z^{+2}e^2\mu^2(1\text{–}3l^2/8r^2)/\{3k_BTr^2\varepsilon(r)\} \tag{127}$$

On the other hand, a modified Lennard-Jones potential for water molecule, $\phi^{ww}(r)$, is useful and is written as follows [45]:

$$\phi^{ww}(r) = 4C\left[(d^w/r)^{12} - (d^w/r)^6\right] - 2\mu^2/r^3 \tag{128}$$

In this equation, the term $4C(d^w/r)^{12}$ is equal to the repulsive part and the parameters C and $d^w$ for water molecule in the gaseous state are equal to 230.9 $k_B$ and 2.824 Å, respectively.

The repulsive part of inter-ionic potential for $\phi^{++}(r)$ may be approximately described as the form of $A^{++}\exp[B^{++}(d^+ + d^+ - r)]$ similar to the repulsive one in Eq. (125), since its interaction occurs around the distance of close contact where the dielectric behavior of neighboring water molecules must be neglected.

Now let us assume that the repulsive potential $\phi_{rep}{}^{+w}(r)$ is represented by the root mean square of $4C(d^w/r)^{12}$ and $A^{++}\exp[B^{++}(d^+ + d^+ - r)]$ as follows:

$$\phi_{rep}{}^{+w}(r) = \left\{4C(d^w/r)^{12} A^{++} \exp\left[B^{++}\left(d^+ + d^+ - r\right)\right]\right\}^{1/2} \tag{129}$$

Insertion of (129) into Eq. (127) gives us the following expression,

$$\phi^{+w}(r) = \left\{4C(d^w/r)^{12} A^{++} \exp\left[B^{++}\left(d^+ + d^+ - r\right)\right]\right\}^{1/2} \\ - z^2e^2\mu^2(1\text{–}3l^2/8r^2)/\{3k_BT\varepsilon(r)r^2\} \tag{130}$$

In a similar way, the inter-particle potential between anion and water molecule is expressed as follows:

$$\phi^{-w}(r) = \left\{4C(d^w/r)^{12} A^{--} \exp\left[B^{--}\left(d^- + d^- - r\right)\right]\right\}^{1/2} \\ - z^2e^2\mu^2(1\text{–}3l^2/8r^2)/\{3k_BT\varepsilon(r^2)\} \tag{131}$$

The dipole moment of water molecule is known to be $\mu = 0.38$ (in the unit of e times 1 Å = $1.6 \times 10^{-29}$ C·m) and l ≒0.5 Å. Therefore, all parameters in (130) and (131) are known. According to Bopp et al. [51], the repulsive parts in (130) and (131) are converted to the exponential decaying functions similar to the repulsive part in (125) [46, 47].

Under these circumstances, it is possible to use either our empirical expressions (130) and (131), or to apply the inter-particle potentials derived by Bopp et al. [51]. It is also possible to estimate the repulsion terms in (130) and (131) by using wave mechanical approach. In fact, the ion-water molecule interactions were obtained by such an elaborating method [33, 52]. However, we will use the above empirical potentials for numerical application, for simplicity.

## 17. Momentum conservation and the tag of water molecules by ion's movement

We will investigate the tag of water molecules by ion's moving in the electrolytic solutions from the view point of equation of motion under an applied field **E** [28].

Under this situation, the second law of motion for the cation i can be written as follows:

$$m^{+}d\mathbf{v}_{i}^{+}(t)/dt = \sum_{j\neq i}^{N+\text{and }N-} \mathbf{f}_{ij} + \sum_{q\neq i}^{X} \mathbf{F}_{iq} + z^{+}e\mathbf{E} \tag{132}$$

$\mathbf{f}_{ij}$ is the force acting on the ion i from the ion j and $\mathbf{F}_{iq}$ is that from the water molecule q.

At the time of steady state, τ, after applying the external field **E**, we have

$$m^{+}\mathbf{v}_{i}^{+}(\tau) = \int_{0}^{\tau} \sum_{j\neq i}^{N+\text{and }N-} \mathbf{f}_{ij}\, dt + \int_{0}^{\tau} \sum_{q\neq i}^{X} \mathbf{F}_{iq}\, dt + \int_{0}^{\tau} z^{+}e\mathbf{E}dt + m^{+}\mathbf{v}_{i}^{+}(0) \tag{133}$$

In a similar way, we have

$$m^{-}\mathbf{v}_{k}^{-}(\tau) = \int_{0}^{\tau} \sum_{l\neq k}^{N-\text{and }N+} \mathbf{f}_{kl}\, dt + \int_{0}^{\tau} \sum_{q\neq k}^{X} \mathbf{F}_{kq}\, dt + \int_{0}^{\tau} z^{-}e\mathbf{E}dt + m^{-}\mathbf{v}_{k}^{-}(0) \tag{134}$$

and

$$m^{w}\mathbf{v}_{q}^{w}(\tau) = \int_{0}^{\tau} \sum_{q\neq(k\text{ and }i)}^{X} (\mathbf{F}_{iq} + \mathbf{F}_{kq})dt + m^{w}\mathbf{v}_{q}^{w}(0) \tag{135}$$

In a unit volume, the total summation of the ensemble averages of these momenta is written as follows:

$$\begin{aligned}
&n < m^{+}\mathbf{v}_{i}^{+}(\tau) > + n < m^{-}\mathbf{v}_{k}^{-}(\tau) > + n_{w} < m^{w}\mathbf{v}_{q}^{w}(\tau) > \\
&\quad = 1/V_{M}\int_{0}^{\tau} \sum (\mathbf{f}_{ij} + \mathbf{f}_{ji})dt + 1/V_{M}\int_{0}^{\tau} \sum (\mathbf{F}_{iq} + \mathbf{F}_{kq} + \mathbf{F}_{qi} + \mathbf{F}_{qk})dt \\
&\qquad + 1/V_{M}\int_{0}^{\tau} (z^{+} + z^{-})e\mathbf{E}dt + n < m^{+}\mathbf{v}_{i}^{+}(0) > + n < m^{-}\mathbf{v}_{k}^{-}(0) > \\
&\qquad + n_{w} < m^{w}\mathbf{v}_{q}^{w}(0) >
\end{aligned} \tag{136}$$

where $n_w$ is the number density of water molecules.

The summation of last three terms on the right hand side of this equation is equal to zero, because there is no external force at $t = 0$. All other terms on the right hand side of this equation are equal to zero by considering the law of action and reaction and charge neutrality condition.

Therefore, we have

$$n < m^{+}\mathbf{v}_{i}^{+}(\tau) > + n < m^{-}\mathbf{v}_{k}^{-}(\tau) > + n_{w} < m^{w}\mathbf{v}_{q}^{w}(\tau) > = 0 \tag{137}$$

This equation indicates that the partial conductivity ratio $< \mathbf{v}_{i}^{+}(\tau)>/ < \mathbf{v}_{k}^{-}(\tau) >$ is not equal to the inverse mass ratio $m^{-}/m^{+}$, which is essentially different from the case of molten salts.

Some of water molecules may be simultaneously pulled by the dissolved ions under an external field **E**. Here, we neglect the relative time relaxation for velocities of particles undergoing the co-operative motion. Taking the numbers of pulled water-molecules by each cation and anion, as $x^{+}$ and $x^{-}$, we have

$$nx^{+} + nx^{-} + x^{r} = n_{w} \tag{138}$$

Here, $x^{r}$ is equal to the number density of un-pulled water molecules.

Since, the movements of remainder water molecules under the external field must be isotropic, we have $x^r < m^w \mathbf{v}^w(\tau) > = 0$. Then $n_w < m^w \mathbf{v}_q{}^w(\tau) >$ is expressed as follows:

$$n_w < m^w \mathbf{v}_q{}^w(\tau) > = x^+ < m^w \mathbf{v}_i{}^+(\tau) > + x^- < m^w \mathbf{v}_k{}^-(\tau) > \tag{139}$$

Insertion of this equation into (66) gives us the following relation:

$$< (m^+ + x^+ m^w)\mathbf{v}^+(\tau) > + < (m^- + x^- m^w)\mathbf{v}^-(\tau) > = 0 \tag{140}$$

Hereafter, we omit the suffix of ion i or k.
Therefore, we have

$$| < \mathbf{v}^+(\tau) > | / | < \mathbf{v}^-(\tau) > | = (m^- + x^- m^w)/(m^+ + x^+ m^w) \tag{141}$$

We cannot apply the above treatment for $H^+$ and $OH^-$ ions, because their conduction mechanisms differ from that of all other dissolved ions. Their mechanisms are known as the Grotthus-type conduction which is a kind of hopping conduction of electrons or holes [3].

It is, however, straightforward to obtain the following relation for all dissolved ions in their dilute limits except for $H^+$ and $OH^-$ ones,

$$| < \mathbf{v}^+(\tau) > | (m^+ + x^+ m^w) = | < \mathbf{v}^-(\tau) > | (m^- + x^- m^w) \tag{142}$$

This relation seems to be valid for all aqueous solutions of equivalent electrolytes in the dilution limit.

Using Eqs. (114) and (115), Eq. (142) for the dilution limit of electrolytic solution is expressed as follows:

$$\begin{aligned}\sigma^+/\sigma^- &= (\text{mass of an anion plus masses of water molecules pulled by its anion})/ \\ &\quad (\text{mass of a cation plus masses of water molecules pulled by its cation}) \\ &= (m^- + x^- m^w)/(m^+ + x^+ m^w) = (m^- < \phi^{-w} >)^{1/2}/(m^+ < \phi^{+w} >)^{1/2}\end{aligned} \tag{143}$$

This equation may correspond to the inverse mass ratio for the partial conductivities of molten salt [6].

## 18. Numerical results in electrolytic solutions

According to the theoretical results we have discussed so far, the pair distribution functions appear in the essential equations [28]. Therefore, how to obtain the pair distribution functions is one of the matters of vital importance.

There are several standard theoretical methods to obtain the pair distribution functions in molecular liquids from the knowledge of inter-particle potentials [33]. In the calculation of site-site distribution function for such a molecular liquid, the reference interaction-site model (RISM) approximation proposed by Chandler and Anderson [52] seems to be useful. Until the present time, the extension of RISM approximation, in order to obtain the potentials of mean force and also the site-site pair distribution functions $g^{\mu\nu}(r)$'s in electrolytic solutions, has been carried out by several authors [53–55]. These attempts cover the insufficient experimental knowledge for pair distribution functions $g^{+-}(r)$, $g^{+w}(r)$ and $g^{-w}(r)$.

However, we will use the $g^{\mu\nu}(r)$'s in aqueous solution of sodium chloride obtained by our own MD simulation. The essential numerical procedure of MD simulation in this study is same as our previous works of molten salts [56]. The procedure of MD simulation of electrolyte aqueous solution will be briefly described as follows for reader's benefit. In MD for the electrolyte aqueous solution, the rigid body models (TIP4P) [57] are used for water molecules. The interactions between constituent TIP4P water molecules are expressed as the charged L-J type potentials, as,

$$_{ij}(r) = \frac{z_i z_j e^2}{r} + \frac{A}{r^{12}} - \frac{B}{r^6} \tag{144}$$

The interactions between alkali metal cation and halide anion, TIP4P- alkali metal anion, and TIP4P – halide anion are expressed as [58]:

$$_{ij}(r) = \frac{z_i z_j e^2}{r} + \frac{C}{r^9} - \frac{D}{r^6} \tag{145}$$

In (144) and (145), i and j stand for the constituent atoms; *e* is the elementary charge. The used charges for the constituent species $z_i$ and the interaction parameters are taken from the literature; TIP4P – TIP4P [57]; TIP4P – alkali metal cation, TIP4P – halide anon, between alkali metal cation, between halide anion, and between alkali metal cation and halide anion [58]. The Ewald method is used for the calculation of the Coulomb interaction. For the structure calculation, MD is performed in NTP constant condition [59–61] under the pressure of 1 atm at 283 K. MD is performed for 50,000 steps with 0.1 fs one time step in 1.1% NaCl aqueous solution. MD cell contains about 10,000 molecules (i.e. 30,000 atoms) for the calculation of the

| Solute | Water (TIP4P) | Cation | Anion |
|---|---|---|---|
| $Li^+$ $Cl^-$ | 10,000 | 112 | 112 |
| $Na^+$ $Cl^-$ | 10,000 | 112 | 112 |
| $K^+$ $Cl^-$ | 10,000 | 112 | 112 |

**Table 2.**
*The numbers of ions in MD cell.*

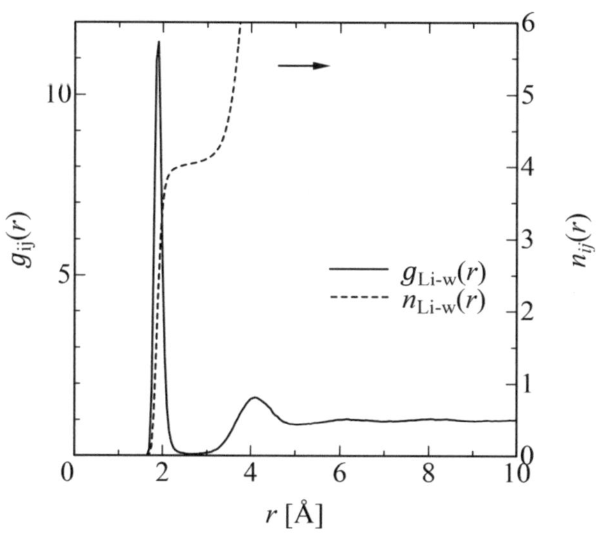


**Figure 5.**
*Pair distribution function of water molecules around a Li + ion, gLi-w(r) in the electrolyte solution of LiCl. And numbers of locating water molecules around a Li + ion within the sphere of the length r centered at its Li + ion, nLi-w(r), in its solution obtained by MD simulation.*

structure and the velocity autocorrelation function for alkali halide aqueous solution. The numbers of the constituent ions in the MD cell are listed in **Table 2**.

The main part of MD is performed using SIGRESS ME package (Fujitsu) at the supercomputing facilities in Kyushu University.

The obtained figures of $g_{ij}(r)$ are shown in **Figures 5–8**. And using these data, we have estimated the numbers of water molecules involved within a sphere of radius r from the centered ion, $n_{ij}(r)$ (i = $Li^+$, $Na^+$, $K^+$ and $Cl^-$; j = oxygen of water molecule) = $4\pi\int_0^r g_{ij}(r)r^2\,dr$, which are also figured in them.

Using Eq. (143), that is, $\sigma^+/\sigma^- = (m^- + x^- m^w)/(m^+ + x^+ m^w)$, and taking an assumption that the pulling water molecules for $Na^+$ ion is equal to 6.0 although its

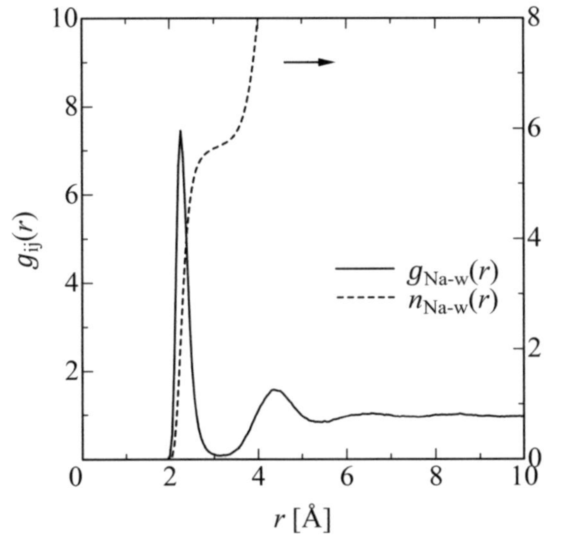


**Figure 6.**
*Pair distribution function of water molecules around a Na + ion, g*Na-w(r) *in the electrolyte solution of NaCl. And numbers of locating water molecules around a Na + ion within the sphere of the length* r *centered at its Na + ion,* nNa-w(r)*, in its solution obtained by MD simulation.*

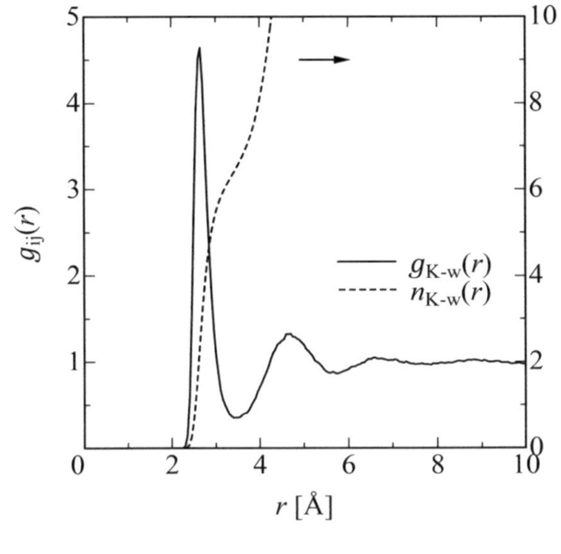


**Figure 7.**
*Pair distribution function of water molecules around a K+ ion, g*K-w(r) *in the electrolyte solution of KCl. And numbers of locating water molecules around a K+ ion within the sphere of the length* r *centered at its K+ ion,* nK-w(r)*, in its solution obtained by MD simulation.*

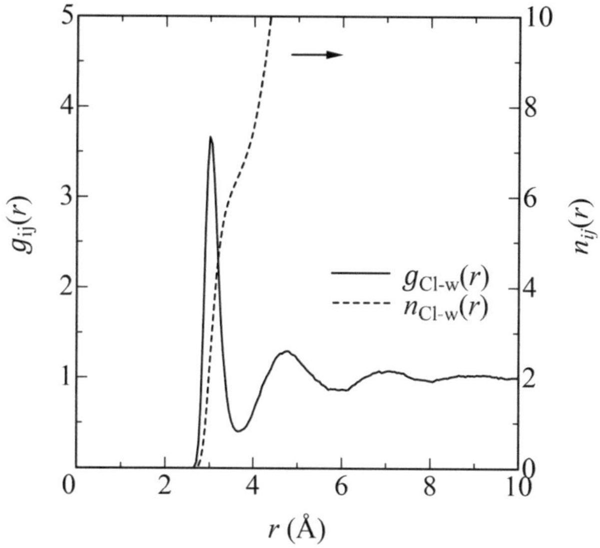


**Figure 8.**
*Pair distribution function of water molecules around a $Cl^-$ ion, gCl-w(r) in the electrolyte solution of MCl (M = Li, Na and K). And numbers of locating water molecules around a $Cl^-$ ion within the sphere of the length r centered at its $Cl^-$ ion, nCl-w(r), in its solution obtained by MD simulation.*

| Ions | Pulling water molecules, $x^+$ or $x^-$ | Hydration numbers in the text book [36] | Hydration numbers obtained from MD simulations |
|---|---|---|---|
| $Li^+$ | 7.6 | 4.3 ± 0.6 | 4.1 |
| $Na^+$ | 6.0* | 5.6 ± 1.7 | 5.7 |
| $K^+$ | 2.8 | 5.5 ± 1.3 | 6.4 |
| $Cl^-$ | 2.8 | 6.0 ± 0.7 | 6.5 |

**Assumption that the pulling number $x^+$ of $Na^+$ ion is equal to be 6.0 and also that the pulling numbers of water molecules for $Cl^-$ are not changed even for that the pairing positive ions are different.*

**Table 3.**
*Numbers of pulling water molecules, $x^+$ or $x^-$ and hydration numbers.*

plausible justification seems to be difficult, then we obtain the pulling water molecules for other ions as shown in **Table 3**, in which the hydration numbers seen in a text book [62] and our results obtained by MD simulation, for reference.

Using these pulling numbers for the constituent ions, we can estimate the term, $(m^- + x^-m^w)/(m^+ + x^+m^w)$ as shown in **Table 4**. As seen in this table, agreements for both terms are satisfactory, which is a kind of proof for the assumption $x^+$ is equal to 6.0.

It is emphasized that the pulling number of water molecules by moving ion has no relation to the hydration number of water molecules as seen in **Table 3**. The hydration of water molecules around electrolytic ions is originated essentially by the thermodynamic stability which is related not only to the interaction energies among ions and water molecules but also to the configuration entropy terms. This is because that the pulling number is not always related to the hydration one.

## 19. Discussion on the electrical conductivities in electrolytic solutions

The present theory seems essentially comparable to the treatments developed by Onsager [19], Fuoss et al. [21], Prigogine [20], Friedman [23], Chandra and Bagchi [27], and Matsunaga and Tamaki [28].

| Electrolyte | $\sigma^+/\sigma^-$ | $(m^- + x^- m^w)/(m^+ + x^+ m^w)$ |
|---|---|---|
| $Li^+ Cl^-$ | 0.595 | 0.598 |
| $Na^+ Cl^-$ | 0.659 | 0.655 |
| $K^+ Cl^-$ | 0.963 | 0.960 |

**Table 4.**
*The ratio of ionic conductivity and the calculation results by using **Table 3**.*

Friedman [23] used a technique of diagram expansion starting from Kubo-Green formula for the conductivity of electrolytic solution and the obtained expression was also written in the form of $\Lambda_c = (\Lambda^+ + \Lambda^-) = \Lambda_0 + \Lambda_1 - kc^{1/2}$. However, his theory is very much sophisticated and too mathematical to understand with a physical insight.

Recent theoretical work carried out by Chandra and Bagchi [27] is basically started from a Kubo-Green type theory, that is, the partial conductivities are derived from velocity correlation functions. Their treatment seems to be a modernized and beautiful and therefore it is very much appreciable. However, the friction force of their theory involves various terms which make it difficult to calculate practically the partial conductivities. In fact, there still remains the task to represent a microscopic formula for $\Lambda_0$.

The present treatment is easily to understand in view of physical insight and is successful for deriving the formula of $\Lambda_0$.

The short-time expansion forms for $< \mathbf{v}_i^+(t)\, \mathbf{v}_i^+(0)>$, $< \mathbf{v}_k^-(t)\, \mathbf{v}_k^-(0) >$ and $< \mathbf{v}_i^+(t)\, \mathbf{v}_k^-(0) >$ are expressed in terms inter-particle potentials and corresponding pair distribution functions as seen in (95) and (97). In the case of molten salts, all these velocity correlation functions yield some physical quantities in relation to a part of partial conductivities [6]. In the present case, however, $Z_\sigma^+(t)$ and $Z_\sigma^-(t)$ play its role. Such an essential difference between the case of molten salt and the electrolytic solution may be ascribed to the difference in the momentum conservation of the system.

## Author details

Shigeru Tamaki[1], Shigeki Matsunaga[2*] and Masanobu Kusakabe[3]

1 Niigata University, Niigata, Japan

2 National Institute of Technology, Nagaoka College, Nagaoka, Japan

3 Niigata Institute of Technology, Kashiwazaki, Japan

*Address all correspondence to: matsu@nagaoka-ct.ac.jp

## References

[1] Janz GJ. Molten Salts Handbook. New York: Academic Press; 1967

[2] Smedley SI. The Interpretation of Ionic Conductivity in Liquids. New York: Plenum Press; 1980

[3] Conway BE. In: Eyring H, editor. Physical Chemistry (an Advanced Treatise) Vol. IX A; Electrochemistry. New York: Academic Press; 1970

[4] Sundheim BR. Transference numbers in molten Salta. The Journal of Physical Chemistry. 1956;**60**:1381

[5] Koishi T, Arai Y, Shirakawa Y, Tamaki S. Transport coefficients in molten NaCl by computer simulation. Journal of the Physical Society of Japan. 1997;**66**:3188

[6] Koishi T, Tamaki S. Partial conductivities of a molten salt based on Langevin equation. Journal of the Physical Society of Japan. 1999;**68**:964

[7] Koishi T, Kawase S, Tamaki S. A theory of electrical conductivity of molten salt. The Journal of Chemical Physics. 2002;**116**:3018

[8] Koishi T, Tamaki S. A theory of electrical conductivity of molten salt II. The Journal of Chemical Physics. 2004; **121**:333

[9] Koishi T, Tamaki S. A theory of transport properties of molten salts. The Journal of Chemical Physics. 2005; **123**(19):194501

[10] Berne BJ, Rice SA. On the kinetic theory of dense fluids, XVI; the ideal ionic melt. The Journal of Chemical Physics. 1964;**40**:1347

[11] Stillinger FH, Lovett R. Ion-pair theory of concentrated electrolytes. I. Basic concept. The Journal of Chemical Physics. 1968;**48**:3858. General Restriction on the Distribution of Ions in Electrolytes. 49, 1991; 1968

[12] Hansen JP, McDonald IR. Statistical mechanics of dense ionized matter. IV. Density and charge fluctuations in a simple molten salt. Physics Review. 1975;**A11**:2111

[13] Giaquinta PV, Parrinello M, Tosi MP. Hydrodynamic correlation functions for molten salts. Physics and Chemistry of Liquids. 1976;**5**:305

[14] Cicotti G, Jaccuci G, McDonald IR. Transport properties of molten alkali halides. Physics Review. 1976;**A13**:426

[15] Trullàs J, Padró JA. Diffusive transport properties in monovalent and divalent metal-ion halide melts: A computer simulation study. Physics Review. 1996;**B55**:12210

[16] Biggin S, Enderby JE. The structure of molten calcium chloride. Journal of Physics C. 1981;**14**:3577

[17] Matsunaga S, Koishi T, Tamaki S. Velocity correlation functions and partial conductivities of molten AgI-AgBr by molecular dynamics simulation. Materials Science and Engineering A. 2007;**449-451**:693

[18] Debye P, Hückel E. The theory of electrolytes. I. Lowering of freezing point and related phenomena. Physikalishce Zeitschrift. 1923;**24**:185

[19] Onsager L. On the theory of electrolytes. II. Physikalische Zeitschrift. 1927;**28**:277

[20] Prigogine I. Non-Equilibrium Statistical Mechanics. New York: Wiley; 1962

[21] Fuoss RM, Accascina F. Electrolytic Conductance. New York: Interscience; 1959

[22] Davis TH, Résibois P. On the relaxation effect in the electrical conductivity of plasmas and electrolytes. The Journal of Chemical Physics. 1964;**40**:3276

[23] Friedman HL. A cluster expansion for the electrical conductance of solutions. Physica. 1964;**30**:509. On the limiting law for electrical conductance in ionic Solutions, 537; 1964

[24] Chandra A, Wei D, Patey GN. The frequency dependent conductivity of electrolyte solutions. The Journal of Chemical Physics. 1993;**99**:2083

[25] Smith DE, Dang LX. Computer simulations of NaCl association in polarizable water. The Journal of Chemical Physics. 1994;**100**:3757

[26] Koneshan S, Rasaiaha JC. Computer simulation studies of aqueous sodium chloride solutions of 298 K and 683 K. The Journal of Chemical Physics. 2000; **113**:8125

[27] Chandra A, Bagchi B. Ion conductance in electrolyte solutions. The Journal of Chemical Physics. 1999;**110**:10024

[28] Matsunaga S, Tamaki S. Ionic conduction in electrolyte solution. Journal of Solution Chemistry. 2014;**43**:1771

[29] Kusakabe M, Takeno S, Koishi T, Matsunga S, Tamaki S. A theoretical extension for the electrical conductivities of molten salts. Molecular Simulation. 2012;**38**:45-56

[30] Tosi MP, Fumi FG. Ionic sizes and born repulsive parameters in the NaCl-type alkali halides—II: The generalized Huggins-Mayer form. Journal of Physics and Chemistry of Solids. 1964;**25**:45

[31] Edwards FG, Enderby JE, How RA, Page DI. The structure of molten sodium chloride. Journal of Physics C. 1975;**8**:3483

[32] March NH, Tosi MP. Atomic Dynamics in Liquids. London: The MacMillan Press Ltd; 1976

[33] Hansen JP, McDonald IR. Theory of Simple Liquids. 2nd ed. New York: Academic; 1986

[34] Mori H. Transport, collective motion, and Brownian motion. Progress in Theoretical Physics. 1965;**33**:423-455

[35] Mori H. A continued-fraction representation of the time-correlation functions. Progress of Theoretical Physics. 1965;**34**:399-416

[36] Copley JRD, Lovesey SW. The dynamic properties of monatomic liquids. Reports on Progress in Physics. 1975;**38**:461-563

[37] Tankeshwar K, Pathak KN, Ranganathan S. Self-diffusion coefficients of Lennard-Jones fluids. Journal of Physics C: Solid State Physics. 1987;**20**:5749

[38] Joslin CG, Gray CG. Calculation of transport coefficients using a modified Mori formalism. Molecular Physics. 1986;**58**:789

[39] Douglass DC. Self Difusion and velocity correlation. The Journal of Chemical Physics. 1961;**35**(1):81

[40] Levesque D, Verlet L. Computer "experiment" on classical fluids III. Physics Review. 1970;**A2**(1970):2514

[41] Tankeshwar K, Pathak K, Ranganathan S. Theory of transport coefficients of simple liquids. Journal of Physics: Condensed Matter. 1990; **2**(1990):5891-5905

[42] Balucani U, Zoppi M. Dynamics of the Liquid State. Clarendon: Oxford; 1994

[43] Hoshino K, Shimojo F, Munejiri S. Mode-coupling analysis of atomic dynamics for liquid Ge, Sn and Na. Journal of the Physical Society of Japan. 2002;**71**(1):119-124

[44] Levenberg K. A method for the solution of certain non-linear problem in least squares. Quarterly of Applied Mathematics. 1944;**2**:164

[45] Berendsen HJC, Griegera JR, Straatsma TP. The missing term in effective pair potentials. The Journal of Chemical Physics. 1987;**9**:6269

[46] Clementi E, Popkie H. Study of the structure of molecular complexes. I. Energy surface of a water molecule in the field of a lithium positive ion. The Journal of Chemical Physics. 1972;**57**:1077

[47] McGlynn SP et al. Introduction to Applied Quantum Chemistry. New York: Rinehart and Winston Inc.; 1972, 1972

[48] Kistenmacher H, Popkie H, Clemnti E. Study of the structure of molecular complexes. V., heat of formation for the $Li^+$, $Na^+$, $K^+$, $F^-$, and $Cl^-$ ion complexes with a single water molecule. The Journal of Chemical Physics. 1974;**59**:5842

[49] Sack H. The dielectric constants of solutions of electrolytes at small concentrations. Physikalishce Zeitschrift. 1927;**28**:199

[50] Oka S. Über den Sattigungszustand einen Dipolflussigkeit in der Umgebung eines Ions. Proceedings of the Physico-Mathematical Society of Japan. 1932;**14**: 441

[51] Bopp P, Dietz W, Heinzinger K. A molecular dynamics study of aqueous solutions X.: First results for a NaCl solution with a central force model for water. Zeitschrift für Naturforschung. 1979;**34a**:1424-1435

[52] Chandler D, Anderson HC. Optimized expansions for classical fluids. II. Theory of molecular liquids. The Journal of Chemical Physics. 1972;**57**:1930

[53] Hirata F, Rossky PJ, Pettitt BM. The interionic potential of mean force in a molecular polar solvent from an extended RISM equation. The Journal of Chemical Physics. 1983;**78**:4133

[54] Pettitt BM, Rossky PJ. Alkali halides in water: Ion-solvent correlations and ion-ion potentials of mean force at infinite dilution. The Journal of Chemical Physics. 1986;**84**:5836

[55] Kovalenko A, Hirata F. Potentials of mean force of simple ions in anbient aqueous solution. I: Three dimensional reference interaction site model approach. The Journal of Chemical Physics. 2000; **112**:10391. II: Solvation structure from the three-dimensional reference interaction site model approach, and comparison with simulation. 10403(2000)

[56] Matsunaga S. Structural features in molten $RbAg_4I_5$ by molecular dynamics simulation. Molecular Simulation. 2013; **39**:119

[57] Jorgensen WL, Chandrasekhar J, Madura JD, Impey RW, Klein ML. Comparison of simple partial functions for simulating liquid water. The Journal of Chemical Physics. 1983;**79**:926

[58] Zhengwei P, Ewig CS, Hwang M-J, Waldman M, Hagler AT. Derivation of class II force field. 4. Van der Waals parameters of alkali metal Cations and halide anions. The Journal of Physical Chemistry. A. 1997;**101**:7243

[59] Andersen HC. Molecular-dynamics simulations at constant pressure and/or temperature. The Journal of Chemical Physics. 1980;**72**:2384

[60] Nosé S. A molecular dynamics method for simulations in the canonical ensemble. Molecular Physics. 1984;**52**:255

[61] Nosé S. A unified formulation of the constant temperature molecular dynamics methods. The Journal of Chemical Physics. 1984;**81**:511

[62] Conway BE. Ionic Hydration in Chemistry and Biophysics. Amsterdam: Elsevier; 1981

Chapter 6

# Study of the Electromagnetic Radiation on the Animal Body

*Leonid Chervinsky*

## Abstract

The rapid technical development of human society on Earth leads to the pollution of its atmosphere and an increase in the electromagnetic radiation of the Sun and its main part—light and ultraviolet radiation. In order to properly protect and control the effects of electromagnetic radiation on the human body, it is necessary to know and understand the process of absorption and conversion of electromagnetic radiation falling on the surface of the body. The material contains the original results of experimental studies on electromagnetic radiation transmission through a sample of quasi-vital skin with pigs of different ages. The reliable results of the percentage ratio of the amount of electromagnetic radiation of the optical spectrum that passes under the skin through the skin layer and the individual wool depending on the species and age of the animal are obtained. The results of the experiment showed that the electromagnetic radiation of the Sun affects the body of the animal through the skin, as well as inside the cylinders of separate wool. This new knowledge is important for biologists and applied engineers to monitor and control electromagnetic radiation for young and old animals with different wools.

**Keywords:** electromagnetic radiation, transmission, reflection, scattering and absorption, skin, animal body surface, conductivity, separate wool

## 1. Introduction

In life, when a person or an animal is under the influence of electromagnetic radiation of the Sun or a special source of electromagnetic radiation of different spectral composition, the biological effect causes radiation energy to enter the surface of its body.

The human body is covered with clothes. The body of the animal is protected by different hairs or wools. It is known that the electromagnetic radiation of the optical spectrum of the short wavelength range (ultraviolet radiation) is very active in the body. Therefore, it is of scientific interest to study the penetration of optical radiation of different wavelengths (spectrum) into the human and animal body. Experimental studies were carried out quickly on the skin with the hair of recently dead animals. The results of such studies are presented below in the described material.

## 2. Methods and installation for studying the paths of penetration of optical and, in particular, ultraviolet radiation into the animal's body

According to the law of geometric optics, the interaction of the optical radiation with the irradiated biological object is characterized by the spectral optical



IntechOpen

absorption coefficients $\alpha(\lambda)$, the reflection $\rho(\lambda)$, and the transmission $\tau(\lambda)$, which are interconnected by the dependence (1):

$$\alpha(\lambda) + \tau(\lambda) + \rho(\lambda) = 1 \tag{1}$$

Measurement of spectral optical characteristics (absorption coefficients $\alpha(\lambda)$, reflection $\rho(\lambda)$, and transmission $\tau(\lambda)$) in the cover of animals is associated with the difficulties of the methodological and technical capabilities in the formation of the experiment.

Firstly, there is a large discrepancy in the magnitude of the coefficients. Based on the analysis of the literature [1, 2] and the results of our own research [3–6], it was found that the magnitude of the reflection coefficients of optical radiation for the shelter of animals is within 0.10–0.60 incident radiation and the transmittance is less than 0.05–0.00001 incident radiation (depending on the thickness of the coating and body pigmentation). Using one method and device measures is impossible without large errors. Therefore, for their measurements, use different methods and different technical devices.

Secondly, the measurement of the optical coefficients of individual skin and wool is associated with similar difficulties due to the large difference in the geometric dimensions of the samples studied: the cross-sectional area of an individual wool has an order of magnitude 0.01 mm$^2$, the area of the skin between the individual hairs is 1 mm$^2$, and the area of irradiation of the skin and wool is regulated by the size of the animal (about 1 m$^2$), and therefore various technical means are used to measure them.

Thirdly, in the system research of the spectrum of optical radiation, there is a need for variable optical systems. In order to increase the flux of the intensity of incident or measured radiation (thus increasing the accuracy of the measurement), the focusing optical systems are used in the studies, which, unfortunately, have different degrees of transparency for different sections of the optical radiation spectrum. So, for the measurement of ultraviolet radiation, optics are used from crystal glass and for visible radiation, from "ordinary" glass, and for infrared radiation, from a special high-temperature quartz glass.

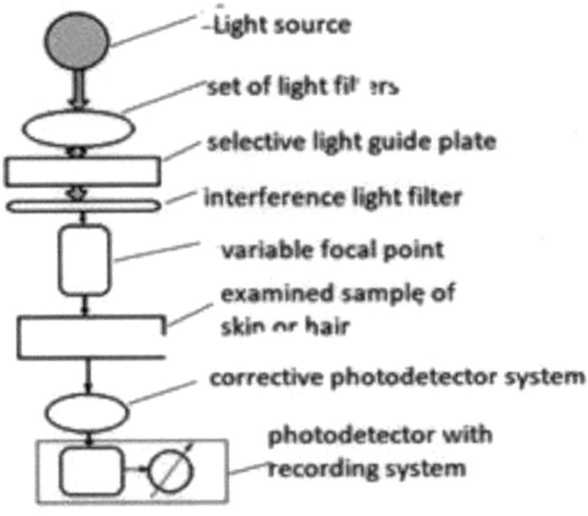


**Figure 1.**
*The block diagram of the installation for the investigation of spectral optical characteristics of the cover of animals.*

Our long-term research in this area has allowed us to develop a method for determining all the named spectral coefficients of the skin and animal wool and the creation of a universal installation for measuring spectral optical coefficients based on a universal fluorescent microscope LUMAM-I3 (Russia) (http://www.laboratorium.dp.ua/item/53).

The structural-optical scheme of the universal device for the study of the optical characteristics of individual wool, areas of clear skin, and areas of the skin with wool is shown in **Figure 1**.

Determination of the spectral coefficients of absorption, reflection, and transmission of the skin and wool cover and its components in animals of different species, age, and breed was summarized on this research facility, made on the basis of the universal luminescent microscope "LUMAM-I3."

The versatility of the proposed installation lies in the fact that the shift of focusing lenses allows you to focus on the irradiation beam (cut from the flow of the optical radiation from the source by a system of quartz and interference filter system) on the plane of the investigated object micro- and macrosize of the area: from 1 $cm^2$ to 0.01 $mm^2$. In this case, the sensitivity of the photodetector by the corrective optical system is regulated in the range from 0.01 $mW/mm^2$ to 1 $W/mm^2$ and takes into account its selective sensitivity to the radiation spectrum. Due to this, it is possible to conduct structural investigations of optical characteristics in the same direction at the same installation:

- Study of optical characteristics of large areas of skin with a wool
- Study of optical characteristics of skin micro areas between individual wools
- Study of optical characteristics and light conductivity of individual wool in animals of all ages and species

It is important that as a result of the study of the optical characteristics and their analysis, it is possible to determine the quantitative scheme of the distribution of electromagnetic radiation parts of the optical spectrum penetrating the skin separately through the skin, through the cylinder separate wool, and as a whole through leather and wool.

In order to increase the reliability of the results obtained in the experiments of measuring the radiation flux from the source was carried out with the help of a photoelectron multiplier, the signal from which was perceived through an amplifier with a sensitive galvanometer, or the corresponding interface through an analog-digital converter was fed for mathematical processing on a computer.

Meanwhile, the measurements of the measuring galvanometer have always been directly proportional to the flux emitted by the photomultiplier (the condition was satisfied so that the intensity of the radiation flux coming into the perimeter window of the photomultiplier was within the linear section of its sensitivity characteristic).

This condition allowed for the calculation of optical coefficients to accurately determine the relative radiation flux through the indicators of the galvanometer, and not in energy units.

In particular, the transmission coefficients of the optical radiation in the investigated samples were determined by the formula (2):

$$\tau_\lambda = \frac{I_\lambda}{I_{O\lambda}} \tag{2}$$

where $I_\lambda$ is the photocurrent in measuring the radiation flux with wavelength $\lambda$ passing through the sample (μA) and $I_{o\lambda}$ is the photocurrent in measuring the radiation flux with wavelength $\lambda$ from the source supplied to the photomultiplier without a sample (μA).

If, in studies, the value of the photocurrent after passing through the sample is much smaller than the photocurrent from the source, $I_{\lambda} \ll I_{o\lambda}$, observed in studies with short-wave ultraviolet radiation at wavelengths less than 365 nm from powerful sources of ultraviolet radiation (due to strong absorption in the irradiation), in the measurement, an error is introduced due to the over-excitation of the photodetector by powerful radiation, and therefore, we have a decrease in its sensitivity (due to the outflow of the linear portion of its h sensitivity characteristics).

To eliminate such an effect, the magnitude of the excitatory radiation from the source was measured by the use of radiation absorbents (nonselective dimming filters of the type NS) with a known attenuation coefficient.

In the described studies, the expression for determining the spectral coefficients of optical transmission has the formula (3):

$$\tau_{\lambda} = \frac{I_{\lambda} \cdot \tau_{\lambda \phi}}{I_{O\lambda \phi}} \tag{3}$$

where $I_{\lambda}$ is the photocurrent in measuring the radiation flux with wavelength $\lambda$ passing through the sample (μA); $\tau_{\lambda\varphi}$ is the transmission coefficient of a neutral filter; and $I_{o\lambda\varphi}$ is the photocurrent in measuring radiation emitted with wavelength $\lambda$ by the photodetector through a dimming filter (μA).

An important optical parameter is also the reflection coefficient of optical radiation from the formula (4):

$$\rho_{\lambda} = \frac{I_{\lambda} - I'_{\lambda}}{I_{\lambda e} - I'_{\lambda}} \cdot \rho_{\lambda e} \tag{4}$$

where $I_{\lambda}$ is the photocurrent in measuring radiation reflected from the sample (μA); $I_{\lambda}$ is the photocurrent in the measurement of radiation, with a wavelength $\lambda$ reflected from the standard (μA); $I'_{\lambda}$ is the extraneous current that shows the galvanometer in the absence of the sample and reference (μA); and $\rho_{\lambda e}$ is the reflection coefficient of radiation with wavelength $\lambda$, for the standard.

A standard used that is commonly accepted in optical studies is the metal plate with a precipitated layer of sulfuric acid barium. The sulfuric acid barium has a comparatively identical reflection coefficient for optical radiation in the range of 640–250 nm equal to 0.96–0.98 [1].

The last important absorption coefficient of optical radiation by the sample was determined by the known formula (1):

$$\alpha_{\lambda} = 1 - \rho_{\lambda} - \tau_{\lambda} \tag{5}$$

It is important to recognize that when measuring the spectral reflection coefficients in the UV region, along with reflected radiation, the radiation of long-wave luminescence of the objects under investigation, which originated from this ultraviolet radiation, was recorded. This radiation made an additional error in the magnitude of the reflection coefficients, but the error introduced did not have a significant effect (given that the intensity of the luminescence was 10n times less than the intensity of the reflected radiation).

The results obtained by this method, the results of the optical characteristics of the skin samples with the wool of different species of animals, and their analysis allow us to determine the quantitative scheme of the distribution of optical radiation penetrating under the skin separately through the skin itself, by individual wool, and are generalized through the leather and wool cover [7–9].

Methods of mathematical statistics are used to process experimental results with given accuracy and reliability. In the processing of experimental data, gross errors were removed from the series. Measurement of optical and electrical quantities was performed at least three times, and the average value of the measured value was analyzed. The variance, coefficient of variation, and mean square deviation of the obtained data were determined.

## 3. Investigation of the light conductivity of individual wool from pigs

In studies devoted to the study of the laws of penetration of optical radiation through the wool coating of animals, the possibility of penetration of radiation under the skin by individual wool, as in fibers, was studied. The main objects were the wool (bristles) of pigs. The choice is based on the fact that pigs' wool has the largest cross-sectional area among other farm animals, and the radiation that passes through the wool cylinder to the photomultiplier tube makes it a valid signal in the linear part of its sensitivity profile.

That is, the value of the photocurrent is greater than the sensitivity threshold of the photoelectron multiplier. This fact increases the reliability of the measurement results. The study scheme is shown in **Figure 2**.

To determine the spectral coefficients of optical conductivity, a system of interference filters with a bandwidth of 10 nm in the range from 300 to 760 nm was used. The corrective lens system is used to ensure uniform irradiation of the sensing area of the sensitive surface of the photo-measuring sensor. To ensure the purity of the study, a rubber seal was applied to the skin from above, in which a separate wool was passed through the hole. The radiation flux was focused on the wool surface above the seal, and the radiation that passed through its cylinder under the skin was recorded.

Then the wool was cut off over the compaction (the radiation was directed perpendicular to the surface of the trimmed hair) and, finally, cut off the follicular

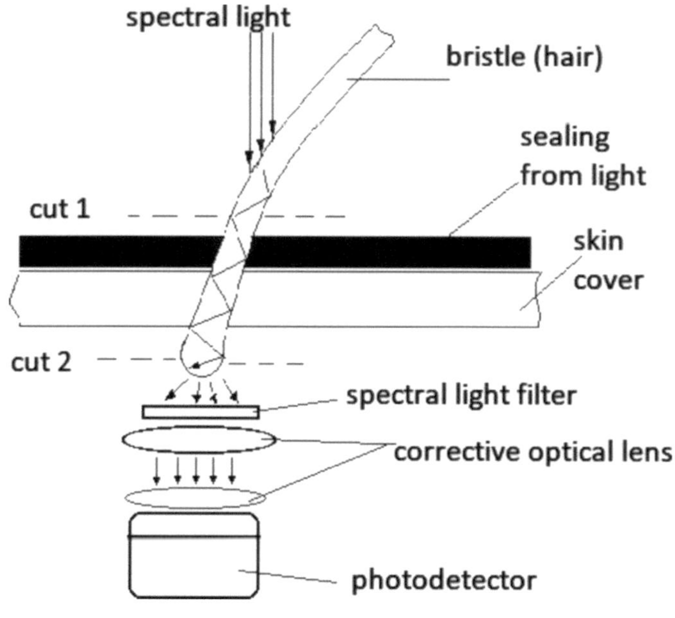


**Figure 2.**
*The structure of the installation for the study of the penetration of electromagnetic radiation of the optical spectrum by a separate cylinder of wool under the skin.*

sphere from the lower part of the skin. Registration of the radiation flux under the skin was carried out in stages, both visually and using a photo-measuring device. **Figure 3** shows the correct results of studying the optical conductivity of individual wool in light pigs (Landrace, Large White).

Analysis of the data shows that white wool conducts up to 10% of the visible electromagnetic radiation falling from above on its surface, depending on the diameter and structure of the wool. When the hair is trimmed over the skin, the radiation emitted into the subcutaneous structure increases. The results of experimental studies of the passage of optical radiation in the cylinder of an individual hair under the skin, on samples of the skin with hair from Large White pigs of different ages, are shown in **Table 1**.

From the tabular data, it can be seen that clipping the sphere of the follicle of an individual wool several times increases the value of the measured radiation. This indicates that the follicle sphere dissipates and absorbs part of the radiation energy (turns into another), that is, the follicle is the particular basis of the photobiological reaction. Practical confirmation of this fact is important for further understanding of the mechanism of penetration of optical radiation into the animal's body and its place of interaction with the biological structures of the skin.

A visual observation of an experiment on the transmission of optical radiation under the skin for different animals (when irradiating a part of the skin surface with wool) also revealed that the follicles have a brighter glow than the inner surface of the skin. Therefore, it is permissible to assert that follicles can be considered as light bulbs in the skin structure and, consequently, primary cells of active photobiological reactions. And there is a realistic explanation: the energy of optical radiation, which reaches the follicles inside the wool cylinder, directly affects the biologically active structures (nerve and blood principles that supply the necessary bioproducts for the formation and growth of wool) and causes a greater photobiological effect than the radiation that passes at the same depth under the skin between individual wool and reaches less active biological entities in the structure of the skin itself [2, 5]. This path is also important because of the fact that the energy of optical radiation enters the wool under the skin directly to the follicle unchanged. In the immediate proximity of the follicle are salogen and sweat glands filled with reactive cellular organic components.

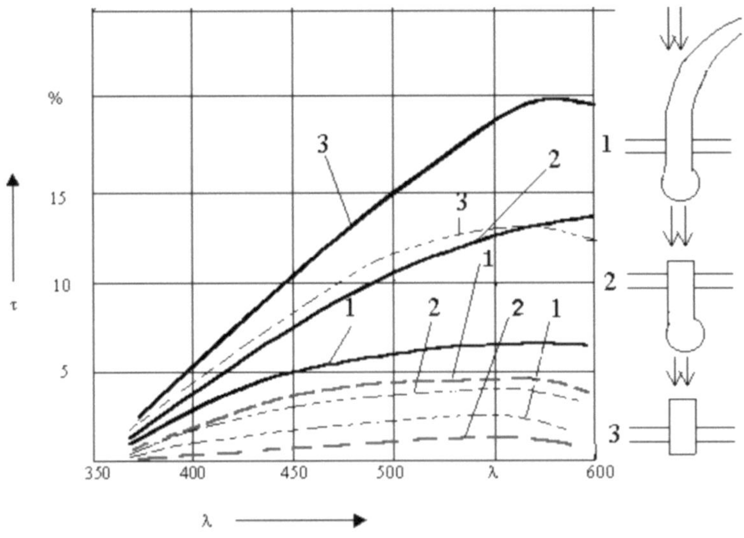


**Figure 3.**
*Spectral coefficients of transmission of a wool and its segments, at a cross section of a wool: __ S = 0.12 mm², _ _ S = 0.08 mm², _ _ S = 0.062 mm².*

| Age of animals | Thickness skin, mm | Cross section of the wool, mm$^2$ | Wool length (to the cut), mm | Follicles | Light conductivity, μA |
|---|---|---|---|---|---|
| 3 years | 3.5 | 0.11–0.13 | Whole wool | Is | 12–14 |
| 3 years | 3.5 | 0.11–0.13 | 2.6–2.8* | Is | 24–28 |
| 3 years | 3.5 | 0.11–0.13 | 2.6–2.8* | The cut | 74–90 |
| 12 months | 2.3 | 0.085 × 0.09 | Whole wool | Is | 9–12 |
| 12 months | 2.3 | 0.085 × 0.09 | 2.6–2.8* | Is | 20–25 |
| 12 months | 2.3 | 0.085 × 0.09 | 2.6–2.8* | The cut | 52–76 |
| 8 months | 2.0 | 0.07–0.074 | Whole wool | Is | 4–7 |
| 8 months | 2.0 | 0.07–0.074 | 2.6–2.8* | Is | 10–12 |
| 8 months | 2.0 | 0.07–0.074 | 2.6–2.8* | The cut | 26–34 |

**The length of the wool segment was measured from the cut above the rubber seal to the surface of the skin.*

**Table 1.**
*The results of experimental studies of the light conductivity of a single wool under the skin of an animal.*

The suitability of this assumption is confirmed by the results of biochemical studies of the effect of optical radiation on the skin, carried out by other authors [3, 5].

To obtain a three-coordinate model describing the spectral dependence of the transmission of electromagnetic radiation on the optical spectrum of a single-cylinder hair on its length, a multiple regression analysis of experimental data was carried out in the Mathcad 2001 Pro software environment. The result of the study is shown in **Figure 4**.

It can be seen from the above that, with the decrease in the wavelength of the optical radiation falling on the surface of a separate one wool, its light conductivity decreases. This corresponds to the results obtained by other authors [2, 4].

The established fact of the propagation of optical radiation energy inside a single-haired cylinder allows determining the intensity of the electromagnetic radiation of the optical spectrum at a specific location of wool (e.g., at the place where it appears above the skin, or at the growth site from the follicle), depending on the distance to the fall site radiation to the surface of a single layer, which is important to know in biological studies when studying the mechanism of the action

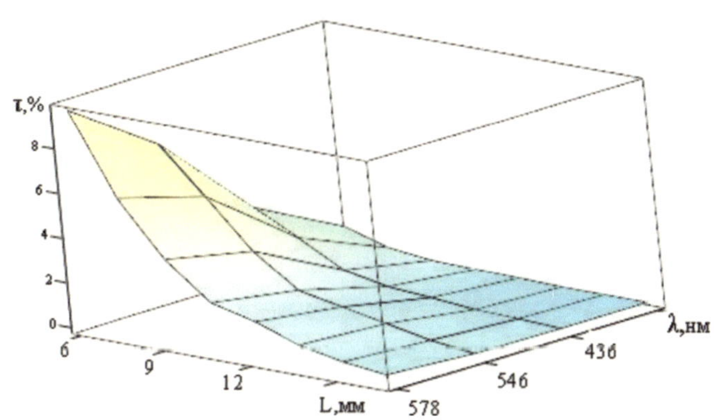


**Figure 4.**
*The dependence of the coefficient transmission of a separate cotton wool τ from the spectral composition of the electromagnetic radiation λ and the length of its light-conducting part L (distance to its follicle).*

of optical radiation through the skin and wools on an animal's body in order to make the desired effective dose and control its effects.

For the visual confirmation of the light conductivity of the wool of different species of animals, the study of the light conductivity of individual wools was carried out by the method of photographing.

The photographs were taken according to the above scheme on **Figure 5**.

**Figure 6** shows an example of a photo of the light transmission of the wool of different animals.

Photo analysis shows that pet hair is mostly transparent to visible radiation. The coat of natural animals is darker with respect to visible radiation. This is natural because wildlife is much more exposed to the open environment and longer exposed

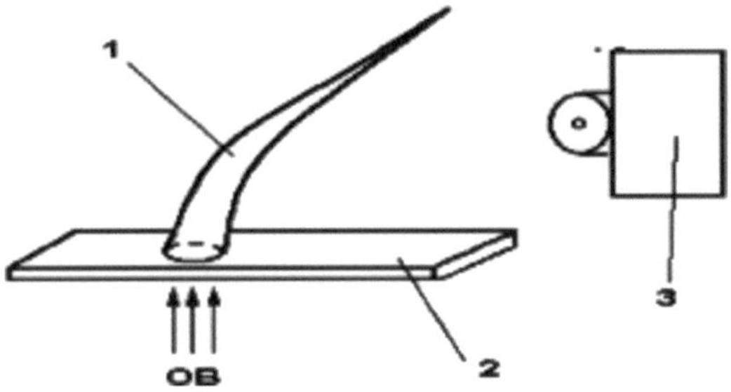


**Figure 5.**
*Scheme for photographing the light conductivity of a single wool. 1, wool; 2, seals; 3, camera; OP, optical radiation.*

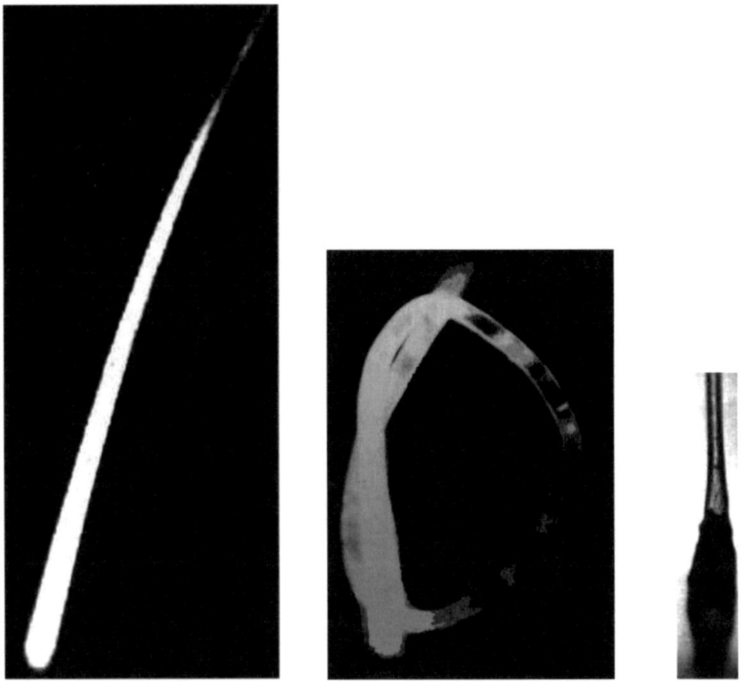

a) b) c)

**Figure 6.**
*Photos of the light conductivity of single wools from different animals: (a) horse tail, (b) wool of pigs, and (c) of the one wools' follicle.*

to sunlight. In summer, when there are long days and intense sunlight, most wild animals have a dark coat color to protect against excessive radiation; in winter, most animals "melt," turning their wool into lighter and thicker ones, which protects from cold and contributes to a wider use of heat when there is less solar radiation in summer.

The calculations presented further confirm that the wool coating plays a significant role in the transfer of the energy of optical radiation of an animal's organism (especially in domestic animals that genetically experience the need for solar radiation under cultivation in a closed space) due to better conductivity and also due to the transfer of energy of this radiation directly to biologically active centers in the skin.

## 4. Investigation of optical characteristics of skin and wool cover animals

Given the biological significance of radiation that reaches subcutaneous body structures directly on individual wool, it should be noted that under natural conditions biological action results in radiation that passes under the skin in any way and reaches active structures. Therefore, it is also important to know what proportion of the energy of optical radiation falling on the surface of the animal passes directly through the skin. For this purpose, complex studies were conducted on samples of the skin with wool, in which the radiation was focused on the areas of the skin between the hairs in accordance with the scheme shown in **Figure 1**.

Based on the results of experimental studies, the generalized curves of the spectral transmittance of optical radiation versus skin thickness for animals of light breeds are shown in **Figure 7**.

For animals listed in **Table 1**, the coefficients of complete transmission of electromagnetic radiation of the visible spectrum of the skin with an area of 1 $cm^2$ were investigated. The three-coordinate **Figure 8** shows, after mathematical processing and generalization of the experimental data, the results of studies of the dependence of the total transmittance of skin samples with wool for the radiation of optical spectral radiation and the thickness of the animal skin.

**Figure 8** shows that in the spectral range of the ultraviolet and visible areas of optical radiation with a decrease in the wavelength of radiation, the intensity and

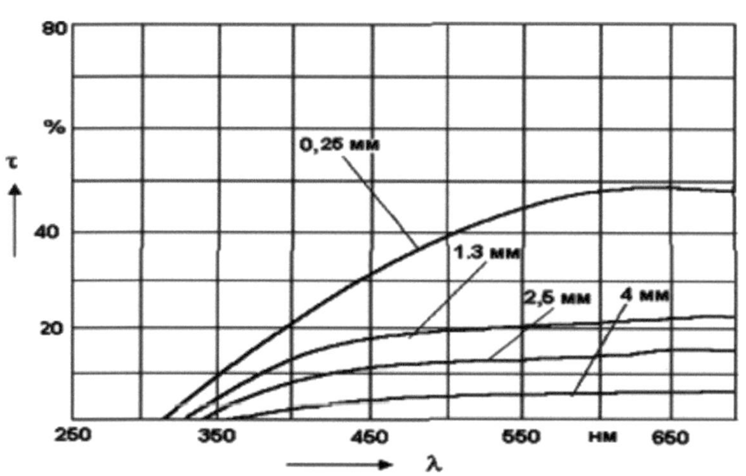


**Figure 7.**
*The transmission coefficient of electromagnetic optical radiation τ through only the skin of agricultural animals of breeds depending on the wavelength λ. Note: the skin thickness is indicated on the curves.*

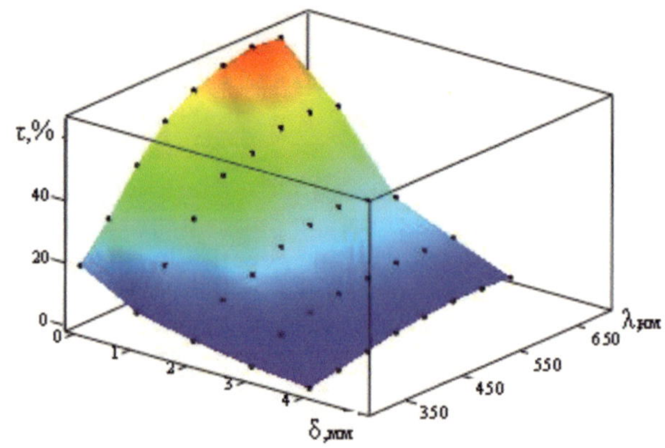


**Figure 8.**
*Three-coordinate picture of the transmission τ of the spectrum of optical radiation to the depth of the skin and wool of the animal, depending on the wavelength of optical radiation λ and the thickness of the cover δ.*

depth of penetration of its energy into the thickness of the leather and wools of farm animals decrease, which corresponds to the results of studies of other scientist experimenters.

## 5. Investigation of the percentage efficiency of radiation penetration through the structure of the coating (through the skin and the wool cylinders) into the animal's body

Establishing the fact of transmission of the electromagnetic radiation of the optical spectrum of an individual wool into the depths of the animal's body, together with the transmission of radiation through the entire skin and coat, is important for understanding and developing research on the mechanism of the action of electromagnetic radiation on the animal and possibly humans. As shown above, electromagnetic radiation on a separate cotton wool with its initial growth (follicle) more effectively affects subcutaneous biological structures than radiation penetrating the skin through the thickness, since the follicle has a well-developed network for the life of the blood vessels and nerve endings.

In the immediate vicinity of the follicle are sebaceous and sweat glands, filled with reactive organic secretions, and it is easy for them to consume the energy of absorbed radiation from the follicles themselves for the development of the skin and body [5].

Although the individual wool in the body of the animal comes a smaller amount of electromagnetic radiation of the optical spectrum than through the skin between the one wool, it comes directly to the biologically active components of the body cells of the animal and can have a greater effect on the photobiological reactions of its development.

In accordance with the foregoing, it is important to know how much of the optical radiation affects the body of an animal through the surface of the skin and how much it passes through individual wools (one wool). To do this, it is necessary to conduct quite complex experimental studies. The difficulty is that it is necessary to irradiate only a certain micro part of the skin surface (or individual wool) and measure the scattered radiation of low intensity that has penetrated deep into the skin.

That is, it is necessary to have a powerful source of optical radiation with a focusing system, and to measure the penetration of the radiation, a photodetector with a perceptual window should be used with dimensions corresponding to the dimensions of the irradiated surface (to reduce the error of measurements from scattering of radiation into skin structures), or to use a complex optical integrated system that further reduces the sensitivity of the instrument (accuracy of measurements). Below is a specific example of the method of comparative analysis of the intensity of the penetration of optical radiation through the skin and on individual wool.

The criterion for the radiation flux passed into the subcutaneous layers of the animal is taken by the photocurrent of the measuring device, since the value of the photocurrent is directly proportional to the intensity of the radiation coming to the receiving window of the photo registration element. Comparison of light conductivity is carried out for a skin area of 1 mm$^2$, and the conditional wool of the same section, 1 mm$^2$. The relative photocurrent values for comparison are determined from formula (6):

$$I_{sk} = \frac{I_{0sk}}{S_{sk}},\ I_w = \frac{I_{0w}}{S_w} \tag{6}$$

where $I_{0sk}$ is the value of the photocurrent from the radiation that passed through the skin to the photodiode's receiving window (impressions of the experiment), $S_{sk}$ is the area of the skin from which the penetrating radiation was recorded, $I_{0w}$ is the magnitude of the photocurrent from the radiation that passed inside the cylinder of a single wool on the photodiode receiving window (impressions of the experiment), and $S_w$ is the area of the cross section of a single wool from which the penetrating radiation was recorded.

The results of research and calculations for comparing the pure transmission of electromagnetic radiation of wool and skin thickness are presented in **Table 2**.

The relative transmittance of electromagnetic radiation of the solar spectrum is determined to understand the change in the quantitative ratio between the penetration of electromagnetic radiation through the skin and through a separate layer of wool during the growth of an animal according to formula (6):

$$k_c = \frac{I_w}{I_{sk}} \tag{7}$$

Based on the data in **Table 2**, a curve was constructed for seeing the change in the relative transmission coefficient of electromagnetic radiation into the animal's body with age, given in **Figure 9**.

| Age, month | Light conductance and geometric parameters of pigs | | | | | | |
|---|---|---|---|---|---|---|---|
| | Skin | | | | Single wool | | |
| | $\delta$, mm | $S_{sk}$, mm$^2$ | $I_{0sk}$, μA | $I_{sk}$, μA/mm$^2$ | $S_w$, mm$^2$ | $I_{0w}$, $10^{-3}$ μA | $I_w$, μA/mm$^2$ |
| 36 | 4 ± 0.2 | 5.0 | 3.8 ± 0.4 | 0.76 ± 0.08 | 0.03 ± 0.006 | 48 ± 3 | 1.6 ± 0.2 |
| 24 | 3 ± 0.4 | 5.0 | 4 ± 0.4 | 0.8 ± 0.08 | 0.025 ± 0.005 | 36 ± 3 | 1.44 ± 0.2 |
| 12 | 2.8 ± 0.3 | 5.0 | 5 ± 0.4 | 1.0 ± 0.08 | 0.016 ± 0.004 | 22 ± 3 | 1.35 ± 0.2 |
| 8 | 1.8 ± 0.2 | 5.0 | 5.8 ± 0.4 | 1.16 ± 0.08 | 0.01 ± 0.003 | 12 ± 3 | 1.2 ± 0.2 |

*Note: δ is the thickness of the skin where the radiation was measured.*

**Table 2.**
*Comparison table of skin transmittance and a separate wool with increasing age in an animal of a Large White breed of pigs.*

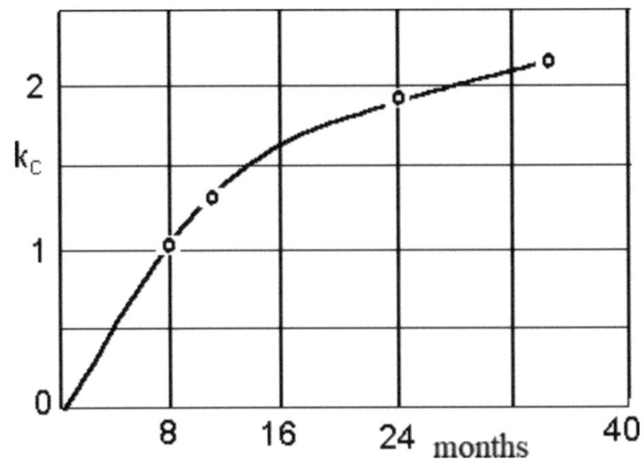


**Figure 9.**
*Change of the relative efficiency of penetration of electromagnetic radiation of the optical spectrum under the skin of an animal by individual wool and through the skin during the growth of the animal (on the example of a White pig).*

From the figure, it is clear that if the animal grows, then the skin becomes coarse and wool cylinders become thicker. This means that older animals have more efficient energy that passes through individual wools.

The analysis of the results of experimental studies shows that the working hypothesis in the process of animal development, the role of wool and skin changes in the interaction of an animal with the environment is real:

1. In young animals, the skin is thin and easily permeable to optical radiation. Therefore, it is natural that the role of the protective screen is woolen, which consists of a thick layer of thin worsted. This layer reduces the flow of optical radiation to the skin by absorbing or dispersing it into the environment and protects the young animal from excessive light energy.

2. With the development of the animal, its skin thickens, its surface and thickness increase, and the penetrating ability of the skin for optical radiation decreases; therefore, the role of the wool as a network of light conductors increases: the distance between them increases on the skin (they are less overlapping each other), their value per unit area of the animal's body decreases, the thickness of each wool increases, and the inner structure of the wool cylinder becomes lighter.

3. Shirting cover with animal growth gradually loses protective functions from optical radiation, acquiring functions, possibly the main conductor of electromagnetic radiation of the optical spectrum in an animal's organism. With the growth of animal's wool become more transparent for optical radiation, which on them as light transmissions with less losses enters the skin to active structures.

4. In real conditions, when an animal is under the radiation of the Sun or a special source of optical radiation of different spectral composition, the biological effect results in all the energy of radiation that has reached the subcutaneous structures of the body by any of these paths and through the skin and individual wools as light conductors.

## Author details

Leonid Chervinsky
National University of Life and Environmental Sciences of Ukraine, Kyiv, Ukraine

*Address all correspondence to: lchervinsky@gmail.com

## References

[1] Smith KC, editor. The Science of Photobiology. NY: Pl. Press; 1977. 430 p

[2] Radha Rani G, Raju GSN. Transmission and reflection characteristics of electromagnetic energy in biological tissues. International Journal of Electronics and Communication Engineering. 2013;**6**(1):119-129. ISSN 0974-2166

[3] Cardona-Hernández Leonel Fierro-Arias MA, Cabrera-Pérez AL, Vidal-Flores AA. Effects of electromagnetic radiation on skin. Dermatologia Revista Mexicana. 2017;**61**(4):292-302

[4] Bashkatov AN, Genina EA, Tuchin VV. Optical properties of skin, subcutaneous, and muscle tissues: A review. Journal of Innovative Optical Health Sciences. 2011;**04**:9-38

[5] Bahar L, Eralp A, Rumevleklioglu Y, Erturk SE, Yuncu M. The effect of electromagnetic radiation on the development of skin ultrastructural and immunohistochemically evaluation with P63. PSP Fresenius Environmental Bulletin. 2018;**27**(3):1764-1771

[6] Chervinsky LS. The action lights on the derma animal's. In: Extra on International Conf. 1st Congress of the World Association for Laser Therapy «WALT»; 1996 May 5-9; Jerusalem, Israel. 1996. p. 2922

[7] Chervinsky LS. Investigation of the light-conductivity of the separate one wool and skins translucence. In: PITTCON'98; 1998 March 1-5; New Orleans, Louisiana, USA. p. 652

[8] Chervinsky LS. About the mechanism of photo reactivation of the biological objects. In: The European Biomedical Optics Week, BIOS Europe'97; 1997 September 4-8; San Remo, Italy. 1997. p. 3198

[9] Chervinsky LS. Primary mechanism of action of optical radiation on living organisms. International Journal of Biosensors & Bioelectronics. 2018;**4**(4):204. DOI: 10.15406/ijbsbe.2018.04.00126

Chapter 7

# Square-Wave Electric Impulses of 10 ms and 100 V/cm of Field Force, Produced by PGen-1 Impulse Generator Device, Affect the Proliferation Patterns of Different Animal Cells

*Bratko Filipič, Lidija Gradišnik, Kristine Kovacs, Ferenc Somogyvari, Hrvoje Mazija and Toth Sandor*

## Abstract

The influence of the medium-strength electric forces (MSE) on the proliferation of adherent chicken embryo fibroblasts (CEF), VERO, MDBK, MRC-5, and HeLa; lymphoblast cells, FB1 and K562; and cell multiplications were analyzed by growth index (GI). Impulse generator device PGen-1 provided 100 V/cm square-wave impulses of 10 ms. Treatment: Samples were subjected to one or three MSE. GIs were compared with controls after 72 hours and one or three treatments: Monolayers: CEF: GI in the control is 16.76, and after one and three MSE, it is 15.81 and 7.09. Vero cells: GI in the controls is 8.39, and after one and three MSE, it is 5.39 and 5.69. MDBK cells: GI in controls is 8.39, and after one and three MSE, it is 5.39 and 5.69. MRC-5 cells: GI in controls is 5.58, and after one and three MSE, it is 4.18 and 2.60. HeLa cells: GI in controls is 13.69, and after one and three MSE, it is 10.16 and 5.37. Suspension cells: Lymphoblast FB1: GI in controls is 6.55, and after one and three MSE, it is 13.48 and 12.25. Lymphoblast K562: GI in controls is 9.07, and after one or three MSE, it is 12.37 and 13.55. To conclude: MSE in monolayer cells inhibits the GI, depending on the nature of cells. MSE enhances the multiplication of lymphoblast FB1 or K562.

**Keywords:** square-wave electric impulses, monolayer cells, lymphoblast cells, growth pattern, growth index decrease, growth index increase, Caspase-3 containing cells, percent of dead cells

## 1. Introduction

Different electric and magnetic field forces can interact with the living systems at enzymatic, cellular, or organism levels [1, 2]. Despite a numerous experimental approaches were performed about this subject, most of the obtained data are completely different and unfortunately very often incomparable. Different basic facts are the reason for this discrepancy. Basically, the living systems *per se* can generate electric or magnetic fields and impulses. These are in the field force range from 10 to 500 mV/cm. The external electric and magnetic range can cause physiological responses at a cellular level. They are of very short forces that can be compared to this duration and are in the range of milliseconds to seconds. Usually they do not result in any important alteration in the system [3, 4]. On the contrary, the biological effects that an electric field with a field force below 1–10 mV/cm causes to a cell are likely due to interaction mechanisms occurring in the cell membrane. For the field forces that are above 1–10 mV/cm, the effects are due to interaction mechanisms occurring in the intracellular compartments. Electric field forces of 10–500 V/cm act as environmental stress factors and result only in a transitory defensive response [5]. In addition, many reports show that these interactions are field force dependent and they can provoke both the enhancement [6] and the inhibition [7] of different cellular functions. It is important that the studied types of responses show the "sensitivity windows" within the above field force range [8]. In the frame of a defined window, the system provides a definite response, while when it is outside the frame, the response disappears or suddenly turns in the opposite direction. The electric field forces that are above 500 V/cm can provoke the sustained response like in the case of electroporation. The most interesting, even the least studied, are the biological effects of medium-strength electric forces of 100 V/cm. The complication, in fact, is that these electric fields can be applied as impulses, long-lasting DC fluxes, or AC waves, and each of them can induce very different types of responses. The applied effects can be a direct ionic current or the electric field generated between the two plates of a condenser, or as a magnetic field, that can provide different interacting doses and different kinds of energies. The named forces can affect the different immune responses of animals [9, 10] or in humans [11] through the induction [12] or augmenting [13] of different immune response elements.

Exogenously added electric impulses can induce the synthesis of antiviral substances, are not interferon, interleukins, or tumor necrosis factors. Therefore, we named them "interferon-like molecules." In our experiments, we also found [14] that medium-strength square-wave impulses of direct ionic current (DC) can result in a short-term direct antiviral resistance to virus infections and in a consequence to alter the membrane properties of the target cells [15]. We therefore decided to study the changes in the expression of membrane-bound surface marker molecules that are on the surface of the immune competent cells in the human blood [16–18]. Our experiments were aimed to establish whether such exogenous electric stimulation of human leukocytes could be utilized as an immune enhancer and an antiviral protector ex vivo, preferably coupled to the dialysis process. In order to detect the potential hazards of such an application, we have to study further the effects of electrostimulation on some other parameters of the human blood.

Least but not the last, the nature and biological conditions of the target system can determine the type and extent of the response. The given conditions can induce an enhancement of cell proliferation in a suspension cell culture and the inhibition of cell proliferation in an adherent growing in monolayer cell culture [19]. Therefore, the herein presented experiments are aimed to investigate the influence and some mechanisms of the medium-strength square-wave electric impulses of the field forces of 100 V/cm, on the proliferation pattern of different animal cells growing in a monolayer or growing as the suspension culture.

## 2. Material and methods

### 2.1 The electric impulse generator device PGen-1

The electric impulse generator device PGen-1 was developed and physically realized by Dr. Sandor Toth and Dr. Ferenc Somogyvari. The PGen-1 device provides 1–300 V/cm square-wave impulses of 1–10 ms duration, with a repetition option of 1–9. It has also a continuous work option. Its repetition intervals can be set between 1 and 10 s. The device consists of two separate circuits: (1) a low-voltage circuit, running on transformed and rectified net current, being stabilized by monolithic integrated stabilizers. They work as a power source for the regulator. (2) A high-voltage circuit serves for the impulse generator itself. The outgoing voltage is regulated by a phase-splitting dimmer and is rectified and stabilized. The low-voltage settings secure the filter condenser. The analogous regulator system is composed from a stable multivibrator, governing the counter, and two synchronized mono-stable multi vibrators, generating the outgoing impulses and the visual control of the signals. The outgoing square-wave impulses are characterized in the **Appendix.** The sample chamber has a 50 ml capacity and is a polypropylene tube with a 2.5 cm in diameter, with a platinum wire electrode. It is detachable from the basic device and is autoclavable. Further details are available in the patent description of the PGen-1 device [20] (**Figure 1**).

### 2.2 The dosimetry

The adsorbed doses (AD in J/g) of the electric impulses in the samples were calculated according to Pakhomov et al. [21] by the use of their formula as follows:

$$AD = \frac{(E^2 \times d^2 \times W \times n)}{(R \times M)} \tag{1}$$

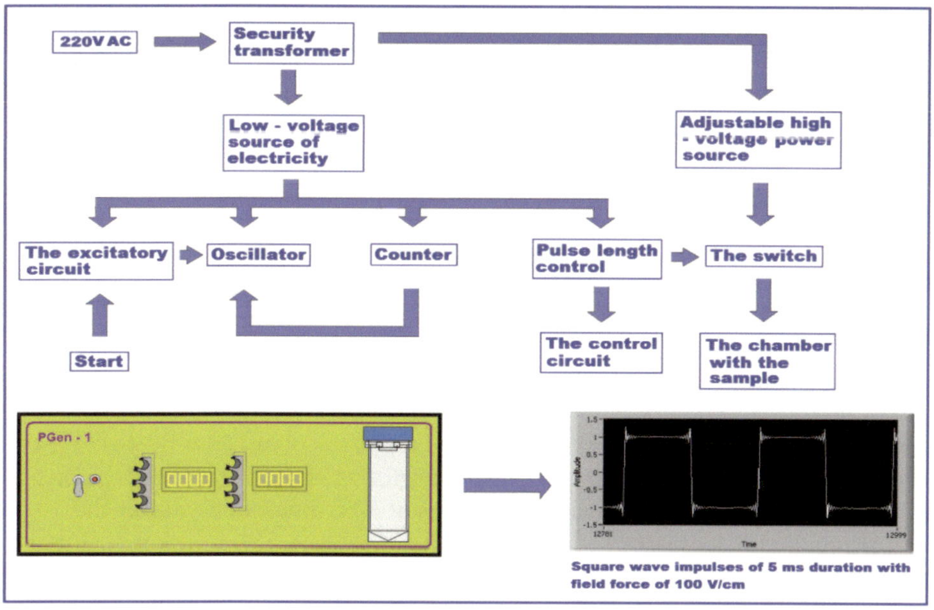


**Figure 1.**
*Electric circuits of the electric impulse generator device PGen-1.*

| Number of impulses | AD in J/g |
|---|---|
| 1 | 0.163 |
| 3 | 0.490 |

**Table 1.**
*The adsorbed dose (AD in J/g) of the electric impulses in the samples.*

In it E is the E is field in the sample (V/m), W is the pulse width ($10^{-8}$ s), n is the number of electric impulses delivered to the sample, d is the gap in the cuvette, R is the resistance of the cuvette with the sample (8–9 Ù), and M is the mass of the medium in g. The AD values in J/g of one and three impulses are shown in **Table 1**.

## 2.3 The cells used in the experiments

The following cells were used: (a) monolayer cells (chicken embryonic fibroblasts, VERO, MDBK, MRC-5, and HeLa cells) and (b) cells growing in suspension (lymphoblast cells like FB1 and K562).

## 2.4 Cell multiplication

The monolayer cell cultures were resuspended in 3 × 50 ml of medium EMEM complemented with surplus $Ca^{2+}$ and 2% of fetal calf serum (FCS). The cell lines growing in suspension were multiplied in the medium RPMI 1640 + 10% FCS. For the experiments they were resuspended in 3 × 50 ml of the medium RPMI 1640 with surplus $Ca^{2+}$ and 2% FCS.

## 2.5 The cell treatments with the electric impulses

The aliquots of 3 × 50 ml of cells suspensions with the $10^5$ cells/mL were put in the three electro-induction chambers with built-in platinum wire electrodes, for treatment. Samples were subjected to one or three square-wave impulses of 10 ms with a field force of 100 V/cm (**Figure 2**). The untreated cells in the electro-induction chamber with built-in platinum wire electrodes represented the cell control.

## 2.6 The growth parameters determination

### *2.6.1 Growth index (GI)*

The GI was determined by the use of spectrophotometer measurements. For the experiments, there were three samples for (1) untreated cell control, (2) one impulse, and (3) three impulses. Cells were cultured in microtiter plates with EMEM medium supplemented with 10% FCS. On the next day, the medium was replaced by new medium containing 5% of FCS. The initial number of cells was determined separately. After 3 h of 5% $CO_2$ incubation at 37°C, the cells were fixed with a 0.25% solution of Glutar aldehyde, and the plates were cooled to 4°C. The cells from untreated control, cells pulsed with one impulse, and cells pulsed with three impulses were incubated for 3 days at 37°C and 5% $CO_2$. After the incubation microtiter plates with cells were fixed with a 0.25% solution of Glutar aldehyde for 20 minutes, washed with phosphate buffer saline (PBS, pH = 7.4), and stained with 4% solution of Methyl blue for 45 minutes at 37°C. Finally, the plates were thoroughly washed with tap water, air-dried, and the colour was extracted by adding of 100 μl of 0.1 M/HCl. The optical density (OD) was measured in AUTOEIA (Lab system) automatic spectrophotometer at 570/650 nm. Every

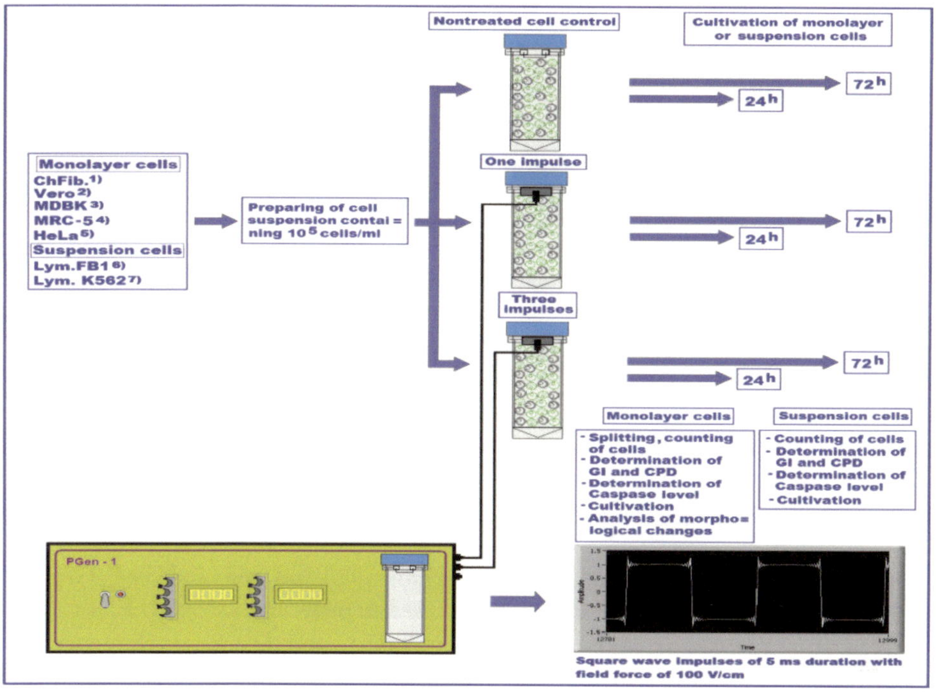


**Figure 2.**
*Scheme of the experiments. (1) chicken embryonic fibroblasts; (2) cercopithecus arthiopis kidney; (3) Madin-Darby bovine kidney epithelial cells; (4) Medical Research Council cell strain five fibroblasts derived from lung tissue; (5) Henrietta lacks cervical tumor immortal cell line; (6) Homo sapiens blood lymphoblast; (7) Homo sapiens bone marrow lymphoblast.*

sample was measured three times and the averages were calculated. The GI was calculated by the following formula: GI = absorbance after 4 days/absorbance of the initial sample [22, 23].

#### *2.6.2 Percentage of GI inhibition*

From the GI of the pulsed cells by the one or three impulses and untreated control cells, the percentage of GI inhibition were calculated by the following formula: percent of GI inhibition = 100 − GI of pulsed cells with one or three impulses/GI untreated control × 100 were calculated [24, 25].

### **2.7 Viability assay**

After the cell's treatment, the monolayer cells were seeded into a 6-well plates containing $3.3 \times 10^5$ cells/well. Twenty-four hours later and 72 h in the parallels, monolayer cells were detached and Trypan blue positive (=blue) cells were counted. The cell numbers were normalized to control cells as 100% [26, 27].

#### *2.7.1 Procedure*

a. Bring adherent cells into suspension by the trypsin/EDTA as described previously, and resuspend them in a volume of fresh medium EMEM at least equivalent to the volume of trypsin. Centrifuge and resuspend the cells that grow in suspension in a small volume of medium. In addition, gently pipette to break up clumps.

b. Under sterile conditions take 100–200 μl of cell suspension.

c. Add an equal volume of Trypan blue (dilution factor = 2) and mix by gentle pipetting.

d. Clean the haemocytometer and the cover slips in 70% ethanol. Clean and dry them with two-site by rubbing with cotton sheets wrapped in cotton cloth.

e. Moisten the cover slip with water or exhaled breath. Slide the cover slip over the chamber back and forth using slight pressure until Newton's refraction rings appear (Newton's refraction rings are seen as rainbow-like rings under the cover slip).

f. Fill both sides of the chamber with cell suspension (approximately 5–10 μl), and view under an inverted phase contrast microscope using the 20× magnification.

g. Count the number of viable (seen as bright cells) and nonviable cells (stained blue). Ideally >100 cells should be counted in order to increase the accuracy of the cell count. Note the number of squares counted to obtain your count of >100.

h. Calculate the concentration of nonviable cells and the percentage of cells using the equation below:

$$\text{non-viable cell count (dead cells/ml)} = \frac{\text{No.of dead cells counted}}{\text{No.of large corner squares counted}} \times \text{dilution factor} \times 10.0000 \qquad (2)$$

## 2.8 Caspase-3 assay

### *2.8.1 Procedure*

a. Caspase-3 activities in pulsed cells were assayed using a commercial method based on fluorochrome-labeled inhibitors of caspases (FLICA, ImmunoChemistry Technologies LLC).

b. The monolayer or suspension cells were pulsed with one or three impulses.

c. One hour after treatment, cells were gently removed from cuvettes and resuspended in a medium EMEM without foetal serum.

d. According to the manufacturer's recommendations, cells were labelled with carboxyfluorescein caspase-3 inhibitors for 1 h at 37°C under 5% $CO_2$ and protected from light as it was previously described [28].

e. Cells were washed with PBS buffer to remove the unbound reagent.

f. Cell fluorescence was determined by the use of fluorescent microscope.

g. Approximately hundreds of cells were acquired for analyses and expressed as percentage of cells showing positive fluorescence.

h. As electric fields were increased, greater numbers of cells became Caspase-3-positive with a homogeneous shift of cells into the Caspase-3-positive gate, making cell percentages with active Caspase-3 the most accurate and meaningful quantification of active Caspase-3 as an apoptosis marker.

### 2.9 Statistical evaluation

For the level of statistical significance determination ($^{*}p < 0.1$, $^{**}p < 0.05$), the T-test was used. All the data are shown as mean value ± standard deviation. The tests were performed in triplicate and each was repeated three to four times.

## 3. Results

### 3.1 The growth parameters, percentages of dead cells, and percentages of Caspase-3 positive cells of different monolayer cells

#### *3.1.1 The growth parameters and percentage of dead cells after 24 h of incubation*

The main effect of the cell's treatment with one or three impulse with the adsorbed dose (AD) of one impulse 0.163 J/g and of three impulses 0.490 J/g was the GI index inhibition expressed in percentage. The detailed results **(Table 2, Figure 3)** show the following:

Chicken fibroblasts: after one impulse 74.41% and after three impulses 57.2%
MDBK cells: after one impulse 31.35% and after three impulses 30.4%
Vero cells: after one impulse 37.2% and after three impulses 24.7%
MRC-5 cells: after one impulse 38.8% and after three impulses 41.9%
HeLa cells: after one impulse 78% and after three impulses 27%

When the percentages of dead cells after one impulse or three impulses were evaluated, the following data were obtained:

Chicken fibroblasts: after one impulse 9.2% and after three impulses 19.6% of dead cells.
MDBK cells: after one impulse 11.2% and after three impulses 12.7%.
Vero cells: after one impulse 8.6% and after three impulses 24.7%.
MRC-5 cells: after one impulse 14.7% and after three impulses 32.6%.
HeLa cells: after one impulse there were 7.4% of dead cells, and after three impulses, there were 11.3% of dead cells.

#### *3.1.2 The growth parameters and percentage of dead cells after 72 h of incubation*

After the cell's treatment with one or three impulses with the adsorbed dose (AD) of one impulse was 0.163 J/g and of three impulses with 0.490 J/g, was the growth index inhibition expressed in percentage. The results presented in **Table 2** and **Figure 4** show the following:

| | | | Monolayer cells | | | | | Suspension cells | |
|---|---|---|---|---|---|---|---|---|---|
| **Cell treatment** | **Time of treatment** | | **Chicken fibroblasts** | **MDBK[1]** | **Vero[2]** | **MRC-5[3]** | **HeLa[4]** | **FB1[5]** | **K 562[6]** |
| Untreated control | 24 h | GI[7] | 1.38 ± 0.12 | 1.19 ± 0.11 | 1.52 ± 0.13 | 1.5 ± 0.13 | 1.73 ± 0.15 | 0.94 ± 0.08 | 1.19 ± 0.11 |
| | | GI inh. % | 0 | 0 | 0 | 0 | 0 | 0 | 0 |
| | | % of dead cells | 4.5 | 3.9 | 5.2 | 6.2 | 5.2 | 4.2 | 6.1 |
| | | % of Casp.3 pos. cells | 2.34 | 2.02 | 2.70 | 3.22 | 2.70 | 2.18 | 3.17 |
| | | CPD[8] | 0.47 ± 0.04 | 1.37 ± 0.12 | 0.23 ± 0.02 | 0.32 ± 0.02 | −0.34 ± 0.03 | −0.1 ± 0.009 | 0.22 ± 0.02 |
| | 72 h | GI | 2.32 ± 0.2 | 2.95 ± 0.2 | 2.77 ± 0.2 | 2.00 ± 0.18 | 2.67 ± 0.24 | 2.78 ± 0.25 | 3.82 ± 0.34 |
| | | GI inh. % | 0 | 0 | 0 | 0 | 0 | 0 | 0 |
| | | % of dead cells | 5.2 | 3.7 | 4.9 | 6.8 | 7.3 | 5.1 | 6.7 |
| | | % of Casp.3 pos. cells | 2.02 | 1.66 | 2.20 | 3.06 | 3.28 | 2.29 | 3.01 |
| | | CPD | 1.21 ± 0.11 | 1.44 ± 0.13 | 1.09 ± 0.09 | 0.73 ± 0.06 | 0.28 ± 0.02 | 1.43 ± 0.12 | 1.9 ± 0.17 |
| 1 impulse | 24 h | GI | 0.35 ± 0.03 | 0.81 ± 0.07 | 0.96 ± 0.08 | 0.91 ± 0.08 | 0.38 ± 0.03 | 1.19 ± 0.1 | 1.9 ± 0.17 |
| | | GI inh. % | 74.4 | 31.5 | 37.2 | 38.8 | 78 | −27 | −59.1 |
| | | % of dead cells | 9.2 | 11.2 | 8.6 | 14.7 | 7.4 | 2.7 | 2.9 |
| | | % of Casp.3 pos. cells | 5.8 | 6.9 | 5.3 | 9.1 | 4.5 | 1.6 | 1.7 |
| | | CPD | −1.49 ± 0.13 | −0.40 ± 0.03 | −0.43 ± 0.03 | −0.38 ± 0.03 | −2.51 ± 0.22 | 0.40 ± 0.03 | 0.89 ± 0.08 |
| | 72 h | GI | 2.19 ± 0.19 | 2.16 ± 0.19 | 1.78 ± 0.16 | 1.5 ± 0.13 | 1.97 ± 0.17 | 5.16 ± 0.46 | 5.21 ± 0.46 |
| | | GI inh. % | 5.71 | 26.7 | 35.7 | 25.1 | 16.1 | −85.6 | −36.4 |
| | | % of dead cells | 34.7 | 12.7 | 11.4 | 38.4 | 11.2 | 3.2 | 1.8 |
| | | % of Casp.3 pos. cells | 23.5 | 8.6 | 7.7 | 26.1 | 7.6 | 2.1 | 1.2 |
| | | CPD | 1.13 ± 0.11 | 0.99 ± 0.09 | 0.45 ± 0.04 | 0.32 ± 0.02 | −0.15 ± 0.01 | 2.47 ± 0.22 | 2.33 ± 0.21 |
| 3 impulses | 24 h | GI | 0.29 ± 0.02 | 0.56 ± 0.05 | 0.88 ± 0.07 | 1.04 ± 0.09 | 0.28 ± 0.02 | 1.35 ± 0.12 | 1.97 ± 0.17 |

*DOI: http://dx.doi.org/10.5772/intechopen.90506*

| Cell treatment | Time of treatment | | Monolayer cells | | | | | Suspension cells | |
|---|---|---|---|---|---|---|---|---|---|
| | | | Chicken fibroblasts | MDBK[1] | Vero[2] | MRC-5[3] | HeLa[4] | FB1[5] | K 562[6] |
| | | GI inh. % | 79.4 | 30.4 | 41.9 | −13.6 | 27.1 | −13.4 | −3.7 |
| | | % of dead cells | 19.6 | 12.7 | 24.7 | 32.6 | 11.3 | 1.3 | 1.7 |
| | | % of Casp.3 pos. cells | 14.5 | 9.9 | 19.3 | 25.5 | 8.2 | 1.02 | 1.3 |
| | | CPD | −1.78 ± 0.16 | −0.92 ± 0.08 | −0.55 ± 0.05 | −0.20 ± 0.01 | −2.95 ± 0.26 | 0.22 ± 0.02 | 0.94 ± 0.08 |
| | 72 h | GI | 0.98 ± 0.08 | 1.12 ± 0.11 | 1.88 ± 0.17 | 0.93 ± 0.08 | 1.05 ± 0.09 | 5.68 ± 0.51 | 5.33 ± 0.48 |
| | | GI inh. % | 57.7 | 48.3 | 31.9 | 37.7 | 46.9 | −10.1 | −2.3 |
| | | % of dead cells | 63.5 | 15 | 38.9 | 70.1 | 14.5 | 2.2 | 1.2 |
| | | % of Casp.3 pos. cells | 44.7 | 11.7 | 30.5 | 55.09 | 11.4 | 1.7 | 0.9 |
| | | CPD | −0.02 ± 0.003 | 0.04 ± 0.005 | 0.53 ± 0.004 | −0.35 ± 0.03 | −1.03 ± 0.11 | 2.33 ± 0.21 | 2.38 ± 0.21 |

[1]*Cercopithecus arthiopis kidney.*
[2]*Madin-Darby bovine kidney epithelial cells.*
[3]*Medical Research Council cell strain five fibroblasts derived from lung tissue.*
[4]*Henrietta lacks cervical tumor immortal cell line.*
[5]Homo sapiens *blood lymphoblast.*
[6]Homo sapiens *bone marrow lymphoblast.*
[7]*GI, growth index.*
[8]*CPD, cumulative population doubling.*

**Table 2.**
*The growth parameters, percentages of dead cells, and percentages of Caspase-3 positive cells.*

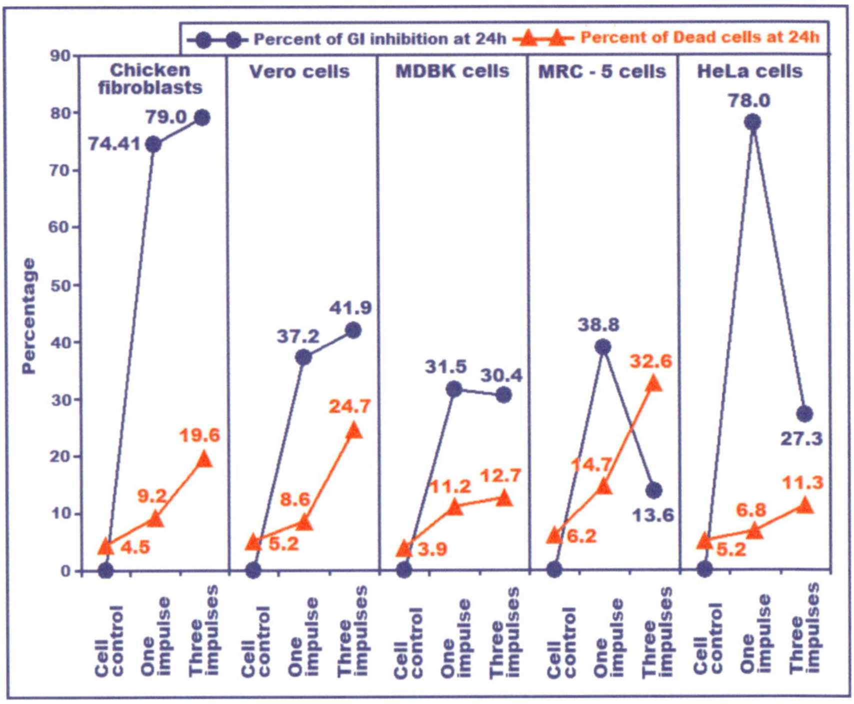


**Figure 3.**
*Growth parameters and percentage of different monolayer dead cells after 24 h of incubation.*

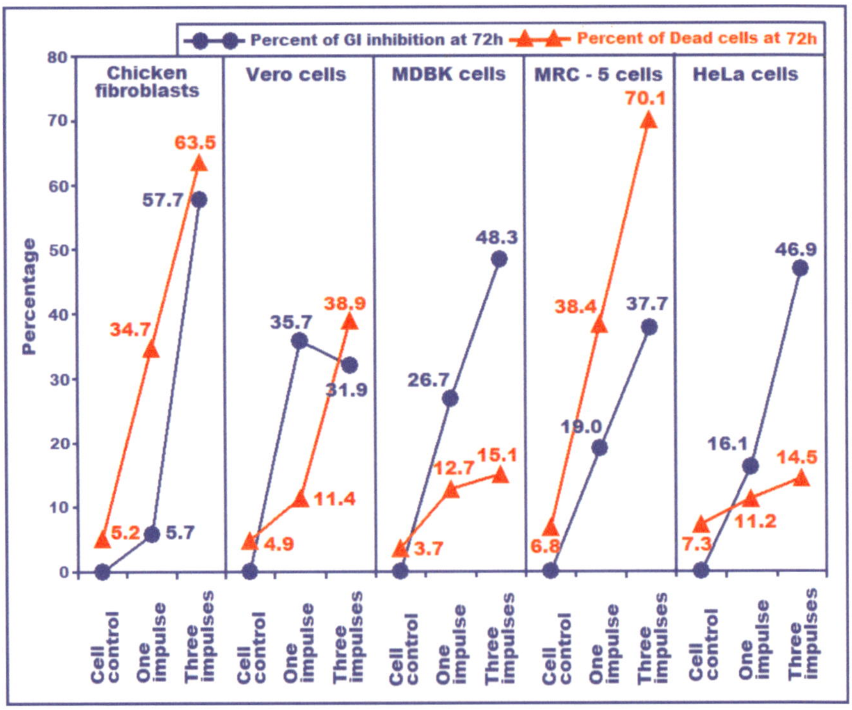


**Figure 4.**
*The Growth parameters and percentage of different monolayer dead cells after 72 h of incubation.*

Chicken fibroblasts: after one impulse 5.7% and after three impulses 57.7%.
Vero cells: after one impulse are 37.7% of GI inhibitions.
MDBK cells: after one impulse 26.7% and after three impulses 48.3%.
MRC-5 cells: after one impulse 25.1% and after three impulses this percentage is 37.7%.
HeLa cells: after one impulse it was 16.1% and after three impulses 46.9%.

The evaluation of the percentages of dead cells after one impulse or three impulses shows the following data for different monolayer cultures: In Chicken fibroblasts the percentage of dead cells after one impulse was 34.7% and after three impulses was 63.5%. In Vero cells after one impulse, there is 11.4% of dead cells and after three impulses 38.9%. In the cells MDBK, after one impulse there is 12.7% of dead cells and after three impulses 15.1%. In the case of MRC-5 cells, after one impulse there is 38.4% of dead cells and after three impulses 70.1%. In the case of HeLa cells, after one impulse there were 11.2% of dead cells, and after three impulses, there were 14.5% of dead cells.

In the evaluation of the trend of GI inhibition percentage, the following comparison was found: In most of the cells, Chicken fibroblasts, MDBK cells, Vero cells, MRC-5 cells, and HeLa cells have higher values after three impulses than after one impulse. The exception is the Vero cells, where the percentage of GI inhibition is a bit lower after three impulses. The analysis about the data of the percentage of dead cells shows the following: The highest percentage of dead cells was 63.5% after three impulses in Chicken fibroblasts. Also was higher percentage of dead cells after three impulses of 38.9. The lowest percentage of dead cells was in MDBK cells after three impulses which were 16.1%. In addition, surprisingly, in the case of MRC-5 cells, again the percentage of dead cells after three impulses is 70.1%. Moreover, the lowest percentage of dead cells was in the case of HeLa cells where it was 14.5%.

#### *3.1.3 Percentage of Caspase-3 positive cells after treating of different cells growing in monolayer, with one or three impulses and cultivated for 24 h or 72 h*

During the experiments, different monolayer cells put into suspension were treated with one or three impulses with the adsorbed dose (AD) of one impulse was 0.163 J/g and of three impulses was 0.490 J/g. After 24 h or 72 h of cultivation, the cells were trypsinized, sedimented by the low speed centrifugation, and analyzed for detection of the percentage of Caspase-3-positive cells (**Figure 5**). In the Chicken embryonic fibroblasts, the percentage of Caspase-3-positive cells after 24 h and one impulse was 23.5% and after three impulses was 14.5%. After 72 h of incubation and one impulse, the percentage of Caspase-3 positive cells was 23.2%. After three impulses, the percentage increased to 44.7%. In the case of Vero cells, after 24 h of incubation and one impulse, the percentage of Caspase-3 positive cells was 7.7%. After three impulses, the percentage increases to 19.3%. After 72 h of incubation and treatment with one impulse, the percentage of Caspase-3 positive cells was 7.5%. After three impulses, this increased to 30.5%. In the case of MDBK cells, after 24 h of incubation and treatment with one impulse, the percentage of Caspase-3 positive cells was 8.5%. After three impulses, this increased to 9.9%. After 72 h of incubation and one impulse, the percentage was 8.6%. After three impulses, this percentage increased to 11.7%. When the MRC-5 cells were analyzed, after 24 h of incubation and one impulse, the percentage of Caspase-3-positive cells was 26.1%; after three impulses, the percentage was 26.5%. After 72 h of incubation and one impulse, the percentage was 26.1%. After three impulses, the percentage highly increased to 55.9%. In the case of HeLa cells after 24 h of incubation and one

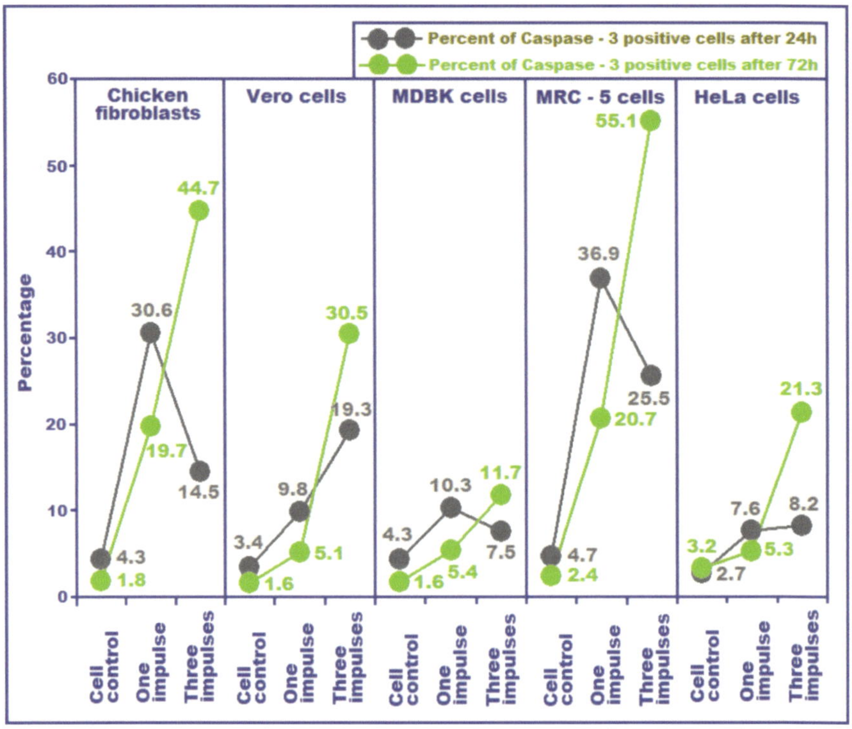


**Figure 5.**
*The percent of Caspase-3-positive cells after 24 h and 72 h of incubation of different types of monolayer cells.*

impulse, the percentage of Caspase-3-positive cells was 7.6%. After three impulses, the percentage was 8.2%. After 72 h of incubation and one impulse, the percentage of Caspase-3 positive cells was 7.6%. After three impulses, this percentage increased to 11.4%.

The comparison of the data on **Figure 5** show the following: (a) The pattern of Caspase-3-positive cells after 72 h and three impulses was the highest in MRC-5 cells 55.1% followed by chicken fibroblasts with 44.7%, Vero cells with 30.5%, MDBK cells with 11.7%, and finally in HeLa cells with 21.3%. (b) The increase of percentage of Caspase-3 positive cells after 72 h of incubation and three impulses was statistically significant in comparison to 24 h of incubation and three impulses. For Chicken fibroblasts, this was 14.5% versus 44.7%. For MRC-5 cells, this was 26.5 versus 55.9%. For Vero cells, this was 19.3% versus 30.5%. (c) For MDBK cells this was 9.9% versus 11.7% and for HeLa cells, this was 8.2% versus 21.3%. For both cell lines, the increase was not statistically significant.

## 3.2 The growth parameters, percentages of dead cells, and percentages of Caspase-3 positive cells of the in suspension growing cells

### *3.2.1 The growth parameters and percentage of dead cells after 24 h of incubation*

After the in suspension growing cell's treatment with one or three impulses having an adsorbed dose for one impulse of 0.163 J/g and for three impulses 0.490 J/g, the main effect was the GI increase expressed in percentage. In comparison, the percentage of dead cells was determined after the treatment with one or three impulses. The detailed results **(Table 2** and **Figure 6)** show the following: For the Lymphoblast cells FB1, the GI increase after one impulse was 27.1%, which fell down to 13.6% after three impulses. In comparison, the percentage of dead cells

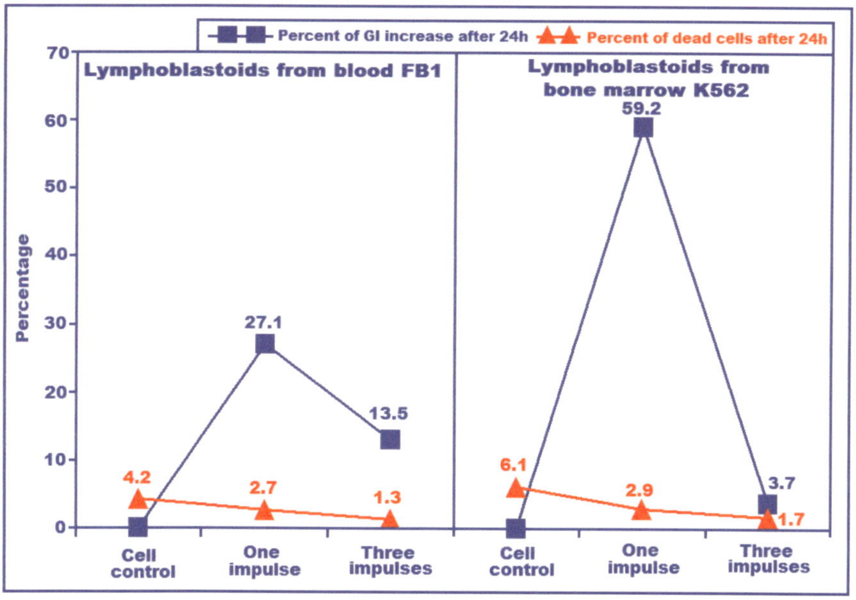


**Figure 6.**
*The Growth parameters and percentage of dead cells after 24 h of incubation and in suspension growing cells treated with one or three impulses.*

after the treatment with one impulse from 4.2% in the control fell down to 2.7%. After three impulses, this percentage fell down to 1.3%. It is unusual that GI increase also fell down after the three impulses.

For the Lymphoblast cells K562, the GI increase was even higher after one impulse, and it was 59.1%. Unexpectedly, after three impulses it fell down to 3.7%. When the percentages after one impulse to 1.7% after three impulses. In general, the GI increase in percentage was higher after one impulse than after three impulses.

### *3.2.2 The growth parameters and percentage of dead cells after 72 h of incubation*

When the cells growing in suspension were treated with one or three impulses with an AD of 0.163 J/g for one impulse and 0.490 J/g for three impulses, after 72 h of incubation (**Figure 7**), the GI increases, which was much higher in the case of lymphoblast cells FB1 after one impulse, which was 86.5%. Again, it fell down to 10.1% after three impulses. In the case of Lymphoblast cells K562, the situation is opposite, namely, after one impulse the percentage fell down to 36.4% and after three impulses even lower to only 2.3%.

The percentage of dead cells shows in lymphoblast cells FB1 the following trend: from the control value of 5.1 to 3.2% after one impulse and to 2.2% after three impulses. In the case of Lymphoblast cells K562, from the xontrol value of 6.7 to 1.8% after one impulse to 1.2% after three impulses. Interestingly, the trend is the same in both cells.

### *3.2.3 Percentage of Caspase-3-positive cells after one or three impulses on suspension lymphoblast cells FB1 or K562 for 24 h*

When the in suspension growing Lymphoblast cells FB1 or K562 were treated with one or three impulses with an AD for one impulse of 0.163 J/g and for three impulses of 0.490 J/g, after the 24 h of incubation were analyzed for percent of Caspase-3 positive cells (**Figure 8**). These values were compared with the

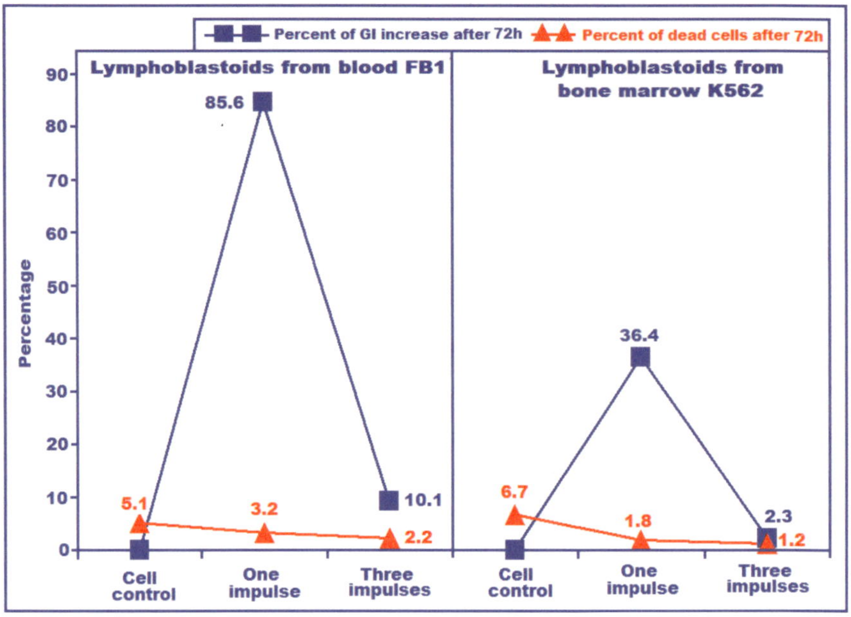


**Figure 7.**
*The Growth parameters and percentage of dead cells after 72 h of incubation and in suspension growing cells treated with one or three impulses.*

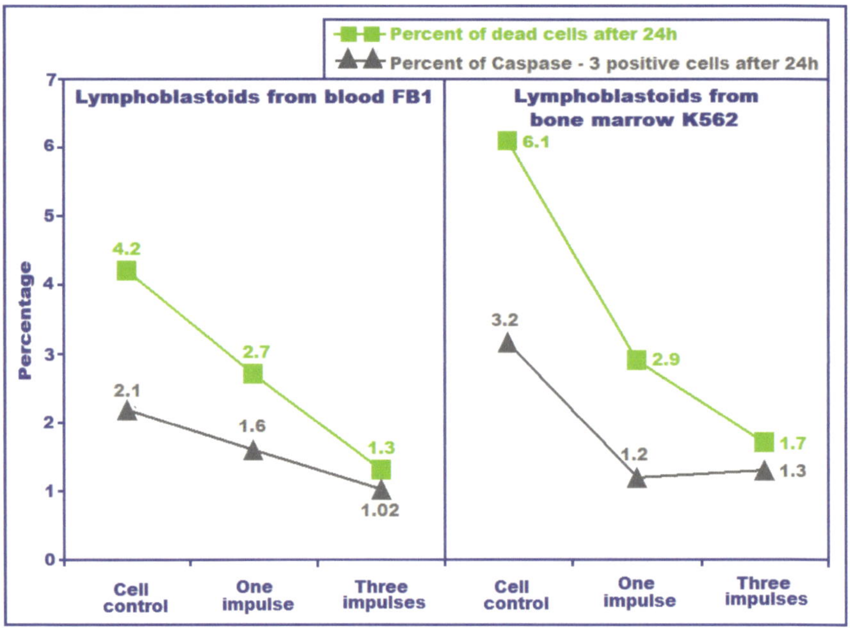


**Figure 8.**
*The Percentage of Caspase-3-positive cells after one or three impulses on in suspension growing lymphoblast cells FB1 and K562 for 24 h.*

percentage of dead cells after 24 h of incubation. In the case of lymphoblast cells FB1 in the cell control, there was 2.18% of Caspase-3 positive cells. After one impulse, this percentage was 2.1%, and after three impulses, it fell down to 1.02%.

The comparison with the percentage of dead cells after 24 h shows the following: In the cell control, there were 4.2% of dead cells. After one impulse, the percentage of dead cells fell down to 2.7%, and after three impulses, it further fell down to 1.3%.

The main conclusion is that the percentage of Caspase-3 positive cells correlates with the percentage of dead cells. In the Lymphoblast cells, K562, in the cell control, there was 3.17% of positive cells. After one impulse, this fell down to 1.2%, and after three impulses this increases a little to 1.2%.

It is interesting that the Lymphoblast cells K562 have one-third higher percentages of dead cells than Lymphoblast cells FB1. In the cell control, there were 6.7% of dead cells. After one impulse, it fell down to 1.8% and after three impulses to 1.2%.

#### 3.2.4 The Percentage of Caspase-3-positive cells after one or three impulses on in suspension growing lymphoblast cells FB1 or K562 growing in suspension, after 72 h

The in suspension growing Lymphoblast cells FB1 or K562 were treated with one or three impulses with an AD for one impulse of 0.163 J/g and for three impulses of 0.490 J/g after 72 h of incubation and were analyzed for the percentage of Caspase-3 positive cells (**Figure 9**). In comparison, the same cells were analyzed for the percentage of dead cells. In the case of lymphoblast cells FB1 in the cell control, there were 4.8% of Caspase-3 positive cells, which fell down to 2.1% after one impulse and further fell down to 1.7% after three impulses. The comparison of the percentage of dead cells shows the following: In the cell control, there were 4.6% of dead cells, which fell down to 1.2% after one impulse and further to 0.9% after three impulses.

The value for the percentage of Caspase-3-positive cells correlates with the percentage of dead cells. The values for the percentage of dead cells were higher than those for Caspase-3. When the lymphoblast cells K562 were analyzed for the percentage of Caspase-3 positive cells, it was found that in the cell control there were 4.6% of positive cells, which fell down to 1.2% after one impulse and to 0.9% after three impulses. The comparison of the percentage of dead cells shows the following: In the cell control, there were 6.7% of dead cells, which fell down to 1.8%

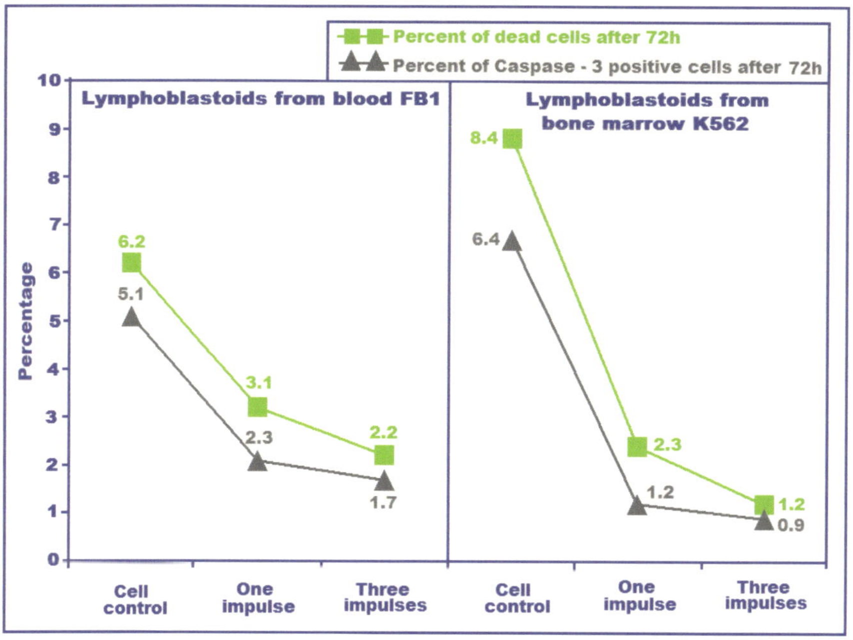


**Figure 9.**
*The Percentage of Caspase-3-positive cells after one or three impulses on suspension lymphoblast cells FB1 and K562 after 72 h of incubation.*

after one impulse and further fell to 1.2% after three impulses. In addition, in this case, the values for the percentage of Caspase-3 positive cells correlate with the percentage of dead cells.

## 4. Discussion

### 4.1 The influence of one or three square impulses with field force of 100 V/cm on different cells growing in monolayer

It was found that nanosecond pulsed electric fields (nsPEFs) can induce the direct electric fields and biological effects on the human colon carcinoma cells [29]. The main effect is the reduction of cell number after electrical impulses, together with the increase of apoptosis markers p53. The dead cells occurred through the necrosis. The direct role of Caspase-3 was not clear. In another research [30], it was found that the picosecond pulsed electric field (psPEF) could induce apoptosis in HeLa cells. The treatments with psPEF led to increased cell apoptosis and cell cycle arrest in the $G_2$/M phase. The psPEF also affected the phosphorylation levels of endoplasmic reticulum sensors and upregulated the expression of glucose-regulated protein 78 (GRP78), glucose-regulated protein 94 (GRP94), and CCAAT enhancer-binding protein (C/EBP) homologous protein (CHOP). All changes were accompanied with the elevation of intracellular $Ca^{2+}$ concentrations. The activation of Caspase-12, Caspase-9, and Caspase-3 led to the release of cytochrome *c* and the upregulation of Bax or the downregulation of Bcl-2, as observed in the HeLa cells. All these data suggest that psPEF is an efficient apoptosis-inducing agent for HeLa cells, which exerts its effects, at least partially, through the endoplasmic reticulum stress and Caspase-dependent signaling pathways. The MTT assay demonstrated that psPEF displayed strong growth inhibitory effects on HeLa cells. The explained cell death and apoptosis by nanosecond pulsed electric fields were found by Beebe et al. [31]. Despite nsPEFs having multiple cellular targets, these studies show that nsPEF affects the cell viability. The cell death depended on the presence of $Ca^{2+}$. When both events occur, cell death can arise.

In our experiments, different monolayer cells being in suspension were treated with one or three impulses having an AD for one impulse of 0.163 J/g and for three impulses of 0.490 J/g. After the treatment with one or three impulses, different monolayer cells were analyzed for the percentage of growth index inhibition, percentage of dead cells, and percentage of Caspase-3 positive cells. In the evaluation of the trend of percentage of GI inhibition, the following comparison was found: In most of the cells, chicken fibroblasts, MDBK cells, Vero cells, MRC-5 cells, and HeLa cells have higher values after three impulses than after one impulse. The exception is the Vero cells, where the percentage of GI inhibition is a bit lower after three impulses. The analysis about the data of the percentage of dead cells shows the following: The highest was the percentage of dead cells (63.5%) after three impulses in Chicken fibroblasts. Also the percentage of dead cells (38.9%) was higher after three impulses. The lowest was the percentage of dead cells in MDBK cells after three impulses (16.1%). Surprisingly, in the case of MRC-5 cells, again the percentage of dead cells after three impulses is 70.1%. And the lowest was the percentage of dead cells in the case of HeLa where it was 14.5%.

The comparison of the data about the percentage of Caspase-3 positive cells shows the following: (a) The pattern of Caspase-3 positive cells after 24 h and three impulses was the highest in MRC-5 cells with 55.9%, followed by Chicken fibroblasts with 44.7%, Vero cells with 30.5%, MDBK cells with 11.7%, and finally in HeLa cells with 11.4%. (b) The increase of percentage of Caspase-3 positive cells after 72 h of incubation and three impulses was statistically significant in comparison to 24 h of incubation and three impulses. For chicken fibroblasts, the percentage was 14.5% versus 44.7%. For MRC-5 cells, the percentage was 26.5% versus 55.9%. For Vero cells, the percentage was 19.3% versus 30.5%. For MDBK cells, the percentage was 9.9% versus 11.7%, and for HeLa cells, the percentage was 8.2% versus 11.4%. For both cell lines, the increase was not statistically significant.

It is important that our data are in quite good agreement with the data of Zhang [32] even he worked on human A375 melanoma cells only. He found that the nsPEFs failed to induce apoptosis of A375 melanoma cells, though it induced necrosis. However, the viability and migration of A375 melanoma cells were significantly inhibited by nsPEFs. It also suppressed the proliferation of A375 melanoma cells by restricting the cells in G0 phase.

### 4.2 The influence of one or three square impulses with a field force of 100 V/cm on the cells that grow in suspensions

The Ren et al. [33] in their experiments studied the influence of nanosecond pulsed electric fields on the activation of intrinsic Caspase-dependent and Caspase-independent cell death in Jurkat cells in vitro. In other world, is the cell dead in Jurkat cells Caspase-dependent or Caspase-independent. Experiments with U937 Lymphoid cell line cultures treated with pulsed electromagnetic field (PEMF) revealed that PEMF decreased the apoptosis of cells [34]. In order to discriminate and clarify mechanisms responsible for cell death after PEMF, cell culture was stimulated with death inducer Puromycin. Puromycin treatment combined with simultaneous single PEMF exposition reduced the death cell rate. Three times PEMF exposition increased the cell viability by about 25%. These data were in agreement with Grassi et al. [35], who found that 50 Hz EMF exposure enhanced proliferation and inhibited Puromycin induced cell death in human Neuroblastoma and rat neuroendocrine cells.

In our experiments, cells growing in suspension, like Lymphoid cells FB1 or K562, were treated with one or three impulses having an AD for one impulse of 0.163 J/g and for three impulses of 0.490 J/g. After the treatment with one or three impulses, different cells growing in suspension were analyzed for the percentage of GI increase, percentage of dead cells, and percentage of Caspase-3 positive cell decrease.

The pattern of the percentage of GI increase after one or three impulses after 24 h at lymphoblast FB1 shows a GI increase after one impulse of 27.1%, which fell down to 13.6% after cell treatment with three impulses. At lymphoblast K562, the picture is opposite: after one impulse the GI increase was 59.1% versus 3.7% after three impulses. The pattern after 72 h at lymphoblast cells FB1 shows 85.6% after one impulse versus 10.1% after three impulses. In lymphoblast cells K562, the picture is opposite: after one impulse 36.4% and after three impulses 2.3% only.

The percentage of dead cells after one or three impulses after 24 h, at Lymphoblast cells FB1, shows the decrease after one impulse from 2.7 to 1.3% after three impulses. The same trend was found in the Lymphoblast cells K562. The decrease of percentage of dead cells was from 2.9% after one impulse to 1.7% after three impulses. After 72 h, we found the same trend: at FB1 cells, the percentage was 3.2% after one impulse versus 2.2% after three impulses. In K562 cells, the percentage was from 1.8% after one impulse to 1.2% after three impulses.

When the pattern of the changes of the Caspase-3 positive cells after one or three impulses was analyzed, the following picture was found after 24 h: in the lymphoblast cells FB1 after one impulse 2.7% which fell to 1.3% after three impulses. It is similar in the picture in the lymphoblast cells K562: from 1.2% after one impulse to

1.3% after three impulses. Therefore, the percentage of Caspase-3 positive cells is independent from the number of impulses.

The percentage of Caspase-3 positive cells after 72 h in the lymphoblast cells FB1 was 2.1% after one impulse and fell down to 1.7% after three impulses. In the case of Lymphoblast cells K562, the percentage was from 1.2% after one impulse to 0.9% after three impulses.

## Acknowledgements

This research was supported by Ivan Čermak, Crodux plin D.o.o., 10000 Zagreb, Croatia, and it was performed via the Project: "Oncolytic Newcastle disease virus in the veterinary medicine' of the Croatian Institute for Experimental and Translational Medicine, Koledinečka 03, 10040 Zagreb, Croatia. Authors are indebted to Tomaž Velnar for English suggestions.

## Conflict of interests

All authors declared no conflicts of interest.

## Appendix

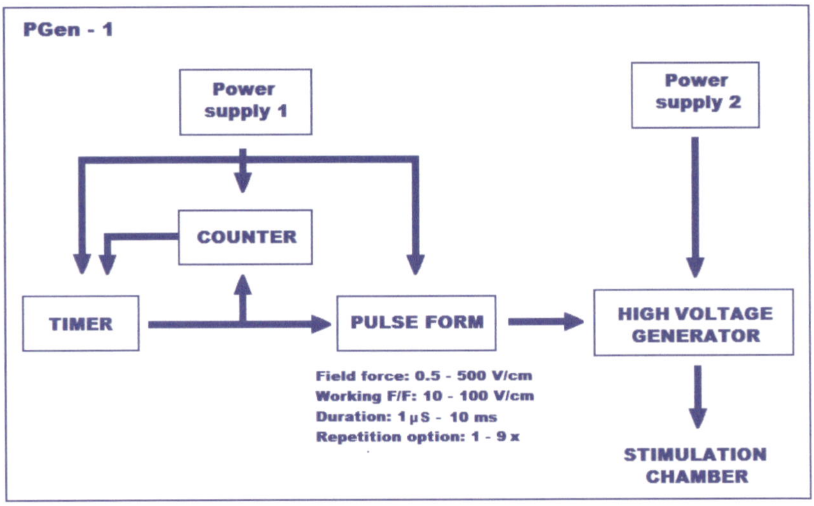


The PGen-1 device generates the characteristic square-wave impulse. This can be mathematically described as Gibbs phenomenon, which is the result of an instantaneous change to a system with a finite number of samples. A pure sine wave has only a single fundamental frequency and no higher-order harmonics. A square wave has an infinite number of frequency components. A sine wave can be transformed into a square wave by adding harmonics to its fundamental. As such, a number of harmonics will produce a perfect square wave with a number of harmonics of 25. Such use of harmonics to build a square wave is demonstrated in the figure below:

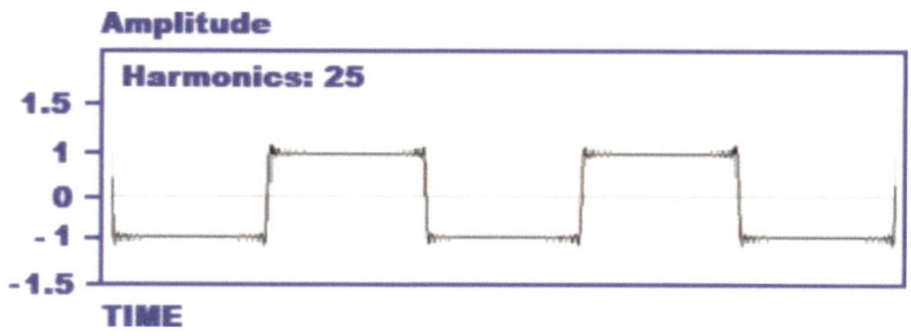


The picture of such a square wave is very similarly with this produced on the bioinduction device described by Pretnar et al. [14] with the number of harmonics of 23.

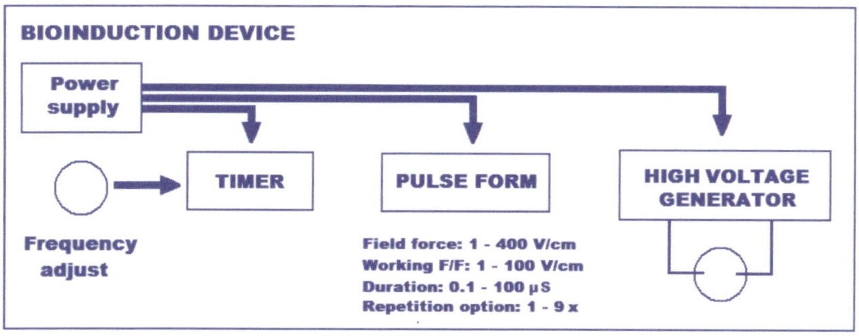


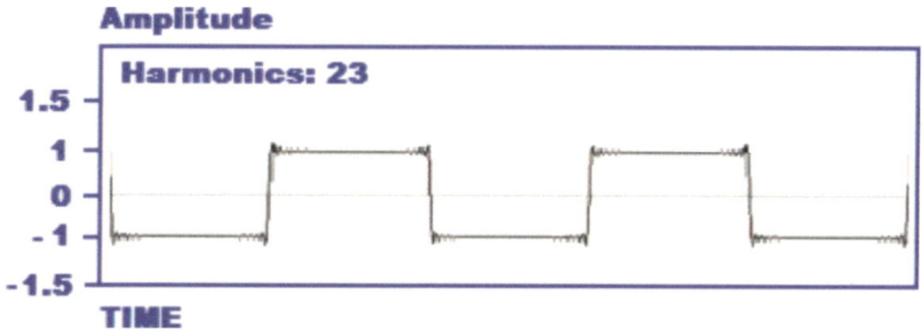

**Author details**

Bratko Filipič[1*], Lidija Gradišnik[2], Kristine Kovacs[3], Ferenc Somogyvari[4], Hrvoje Mazija[1] and Toth Sandor[5]

1 CIETO (Croatian Institute for Experimental and Translational Oncology), Zagreb, Croatia

2 Medical Faculty of Maribor, Institute of Biomedical Sciences, Maribor, Slovenia

3 Sylvester Comprehensive Cancer Center, Coral Glades, University of Miami, USA

4 Department of Medical Microbiology and Immunobiology, University of Szeged, Szeged, Hungary

5 TEWA Consultant, Szeged, Hungary

*Address all correspondence to: bratko.filipic@gmail.com

## References

[1] Frey HA. On the nature of electromagnetic fields interactions with biological systems. FASEB Journal. 1993; 7:272-281. DOI: 10.1096/fasebj.7.2. 8440406

[2] Meulenberg WJC, Todorović V, Čemažar M. Differential cellular effects of electroporation and electrochemotherapy in monolayers of human microvascular endothelial cells. PLoS One. 2012;**7**(12):e52713. DOI: 10.1371/journal.pone.0052713

[3] Bolognani L, Del Monte V, Francia F, Costato M. Low-frequency electromagnetic pulsed field stimulation of yeast. Electromagnetic Biology and Medicine. 2009;**11**(1):1-10. DOI: 10.3109/15368379209012847

[4] Im C-H, Jun CS, Sekihara K. Recent advances in biomagnetism and its applications. Biomedical Engineering Letters. 2017;**7**:183. DOI: 10.1007/ s13534-017-0042-3

[5] Cui Y, Liu X, Yang T, Mei YA, Hu C. Exposure to extremely low-frequency electromagnetic fields inhibits T-type calcium channels via AA/LTE4 signaling pathway. Cell Calcium. 2014;**55**(1): 48-58. DOI: 10.1016/j.ceca.2013.11.002

[6] Araújo QFO, Alice Z, Coelho ZAM, Isabel CP, Margarit CPI, Vaz-Junior AC, et al. Electrical stimulation of *Saccharomyces cerevisiae* cultures. Brazilian Journal of Microbiology. 2004; **35**(1–2):97-103. DOI: 10.1590/ S1517-83822004000100016

[7] Holan V, Machala O, Dao T, Minowada J. Effect of direct current on T cell activity: Modulation of interleukin 2 production. Immunobiology. 1994;**190**:368-371. DOI: 10.1016/S0171-2985(11)80608-X

[8] Berg H. Problems of weak electromagnetic field effects in cell biology. Bioelectrochemistry and Bioenergetics. 1999;**48**(2):355-360. DOI: 10.1016/s0302-4598(99)00012-4

[9] Huang J, Ye Z, Hu X, Lu L, Luo Z. Electrical stimulation induces calcium-dependent release of NGF from cultured Schwann cells. Glia. 2010;**58**(5):622-631. DOI: 10.1002/glia.20951.w

[10] Velizarov S. Electric and magnetic fields in microbial biotechnology: Possibilities, limitations, and perspectives. Electromagnetic Biology and Medicine. 1999;**18**(2):185-212. DOI: 10.3109/15368379909012912

[11] Funk WHR, Monsees KT. Effects of electromagnetic fields on cells: Physiological and therapeutical approaches and molecular mechanisms of interaction. Cells, Tissues, Organs. 2006;**182**(2):59-78. DOI: 10.1159/ 000093061

[12] Petrini M, Polidori R, Ambrogi F, Vaglini F, Zanioli P, Ronca G, et al. Effect of different low-frequency electromagnetic fields on lymphocyte activation: At which cellular level? Electromaganetic Biology and Medicine. 1990;**9**:159-166. DOI: 10.3109/ 15368379009119803

[13] Katsuki S, Li Y, Miyakawa D, Yamada YR, Onishi N, Lim S. Response of mammalian cells to non-thermal intense narrowband pulsed electric fields. In: 11th European Conference of Antennas and Propagation (EUCAP). 2017. DOI: 10.23919/EuCAP.2017.7928345

[14] Pretnar G, Filipic B, Golob A, Skodic A, Toth S, Mecs I, et al. Electroinduction of interferon-like proteins. Bioelectrochemistry and Bioenergetics. 1991;**25**:183-193. DOI: 10.1016/0302-4598(91)87002-X

[15] Filipič B, Kovacs K, Somogyvari F, Ihan A, Ocosovsky I, Koren S, et al. The effects of medium-strength electric

impulses on human blood. Bioelectrochemistry. 2000;**52**(1):29-36. DOI: 10.1016/S0302-4598(00)00079-9

[16] Gal G, Toth M, Toth S, Kiss E. Alpha and gamma interferon production of dialysed chronic uraemic and kidney transplant patients. Abstracts EDTA Brussels Belgium VI. 1985;**25-29**:49-49

[17] Filipič B, Koren S, Kovacs K, Somogyvari F, Toth S. Naprava in postopek za elektrostimulacijo celic (Device and method for electrostimulation of cells) (in Slovenian). Slovenian Patent No. SI 20111 A, 600–214/98-MZT-6 z dne: 03.07. 2000. (P-9800214. 1998)

[18] Kovacs K, Somogyvari F, Filipič B, Mecs I, Toth S. Induction of antiviral response by electric pulses. Electromagnetic Biology and Medicine. 2000;**19**(3):279-287. DOI: 10.1081/JBC-100102119

[19] Filipic B, Gradišnik L, Somogyvari F, Toth S, Mazija H. The biological activity of medium-strength square-wave electric impulses on the proliferation patterns of different animal cells. ALTEX Proceedings. 2019; **8**(1):52-52. DOI: 10.1314/RG.2.2.1938040326

[20] Filipič B, Koren S, Kovacs K, Somogyvari F, Toth S. Naprava in postopek za elektrostimulacijo celic—Device and method for electro stimulation of cells (in Slovenian). Slovenian Patent No. SI 20111A, 600–214/98-MZT-6 z dne: 03.07. 2000 (P-9800214. 1998)

[21] Pakhomov AG, Phinney A, Ashmore J, Walker K III, Kolb FJ, Kono S, et al. Characterization of the cytotoxic effect of high-intensity, 10-ns duration of electrical pulses. IEEE Transactions on Plasma Science. 2004;**32**(4):1579-1586. DOI: 10.1109/TPS.2004.831773

[22] Rink H, Vornhagen R, Koch HR. Rat lens epithelial cells in vitro. I. Observation on aging, differentiation and culture alternations. In Vitro. 1980; **16**(1):15-19. DOI: 10.1007/bf02618195

[23] Filipič B, Golob A, Toth S, Mecs I, Beladi I, Likar M. Interactions between human and porcine interferons. Acta Virologica. 1991;**35**:19-26

[24] Filipič B, Schauer P, Suhar A. Changes of the intracellular levels of some hydrolases during the spontaneous transformation of rat embryonal fibroblasts (REF) (Wistar strain) cell line. Farm. Vest. 1984;**35**(3):239-243

[25] Ito M. Microassay for studying anticellular effects of human interferons. Journal of Interferon Research. 1984;**4**(4):603-608. DOI: 10.1089/jir.1984.4.603

[26] Singh A, Kakkar P, Pant AB. Comparative evaluation of cytotoxic effects of MTAD and sodium hypoclorite using LDH and Trypan blue assays: An in vitro study. Saudi Endodontic Journal. 2018;**8**(3):189-195. DOI: 10.4103/sej.sej_75_17

[27] Jauregui OH, Hayner TN, Driscoli LM, Williams-Holland R, Lipsky MH, Galleti MP. Trypan blue dye uptake and lactate dehydrogenase in adult hepatocytes—Freshly isolated cells, cell suspensions and primary monolayer cultures. In Vitro. 1981; **17**(12):1100-1109. DOI: 10.1007/bf02618612

[28] Ford WE, Ren W, Blackmore PF, Schoenbach KH, Beebe SJ. Nanosecond pulsed electric fields stimulate apoptosis without release of pro-apoptotic factors from mitochondria in B16f10 melanoma. Archives of Biochemistry and Biophysics. 2010;**497**:82-89. DOI: 10.1016/j.abb.2010.03.008

[29] Hall HE, Schoenbach HK, Beebe JS. Nanosecond pulsed electric fields (nsPEF) induce direct electric field effects and biological effects on human

colon carcinoma cells. DNA and Cell Biology. 2005;**24**(5):283-291. DOI: 10.1089/dna.2005.24.283

[30] Chen W-J, Xiong Z-A, Zhang M, Yao C-G, Zhao Z-Y, Hua Y-Y, et al. Picosecund pulsed electric fields induce apoptosis in HeLa cells via the endoplasmic reticulum stress and caspase-dependent signaling pathways. International Journal of Oncology. 2013; **42**:963-970. DOI: 10.3892/ijo.2013.1774

[31] Beebe JS, Sain MN, Ren W. Induction of cell death mechanisms and apoptosis by nanosecond pulsed electric fields (nsPEFs). Cell. 2013;**2**:136-162. DOI: 10.33390/cells1010136

[32] Zhang B. Effect of low-field high-frequency nsPEFs on the biological behaviors of human A375 melanoma cells. IEEE Transactions on Biomedical Engineering. DOI: 10.1109/TBME. 2017.2784546 TBME-01034-2017.R1

[33] Ren W, Sain MN, Beebe J. Nanosecond pulsed electric fields (nsPEFs) activate intrinsic caspase-dependent and caspase-independent cell death in Jurkat cells. Biochemical and Biophysical Research Communication. 2012;**421**:808-812. DOI: 10.1016/j. bbrc.2012.04.09

[34] Kaszuba-Zwoinska J, Wojcik K, Bereta M, Zombier A, Pierzchalski P, Rokita E, et al. Pulsating electromagnetic field stimulation prevents cell death of puromycin treated U937 cell line. Journal of Physiology and Pharmacology. 2012; **61**(2):201-205. DOI: 10.1515/cmble-2015-0006

[35] Grassi C, D'Ascenzo M, Torsello A, Martinotti G, Wolf F, Cittadini A, et al. Effects of 50 Hz electromagnetic fields on voltage-gated $Ca^{2+}$ channels and their role in modulation of neuroendocrine cell proliferation and death. Cell Calcium. 2004;**35**(4):307-315. DOI: 10.1016/j.ceca.2003.09.001